Lecture Notes in Mathematics

A collection of informal reports and seminars
Edited by A. Dold, Heidelberg and B. Eckmann, Zürich

187

Harold S. Shapiro

Department of Mathematics
The University of Michigan, Ann Arbor, MI/USA

Topics in Approximation Theory

Springer-Verlag
Berlin · Heidelberg · New York 1971

AMS Subject Classifications (1970): 41-XX, 42 A04, 42 A08, 42 A64 – 42 A72, 42 A88, 42 A96, 44 A35, 46 B99, 46 C10, 46 E15, 46 E20, 26 A72, 30 A31, 30 A38, 30 A76, 30 A80

ISBN 3-540-05376-X Springer-Verlag Berlin · Heidelberg · New York
ISBN 0-387-05376-X Springer-Verlag New York · Heidelberg · Berlin

Offsetdruck: Julius Beltz, Weinheim/Bergstr.

PREFACE

These notes are based on a course given by me at the Royal Institute of Technology in Stockholm in the Fall Term, 1969, and on some seminar talks I gave in the Stockholm area. In addition, two chapters (presented as appendices) were kindly contributed to the present volume, at my request, by Jan Boman and Torbjörn Hedberg respectively.

The audience at my lectures consisted largely of research scholars from the Mittag-Leffler Institute with strong backgrounds in analysis, and therefore I took for granted some familiarity with measure theory, distribution theory, Fourier analysis, functional analysis and even more specialized topics like H^p spaces and vector-valued integration. On the other hand, no prior knowledge of approximation theory was assumed. In writing up the notes, I tried to smooth the path of the beginner by giving references for those notions not likely to be familiar; I did not always do this, however, for fairly standard topics of real, complex, or functional analysis (e.g. the Hahn-Banach theorem) since good text-books on these subjects are legion.

It should be stressed that these notes are not meant to serve as a text-book for a course in approximation theory, although portions of it might advantageously be used in connection with such a course or seminar. The choice of topics is somewhat arbitrary, the coverage of these is mostly not systematic, and there is considerable disparity in level and style between different chapters. (Thus, for example, there is no discussion of such fundamental topics as splines, polynomial and rational approximation in the complex domain, or algorithms for constructing best approximations.) It was my intention that this book should be used in conjunction with existing works, and I have frequently referred to the literature rather than repeat demonstrations in readily available sources. The exercises also vary widely in character; a serious reader should not just skip over them, since occasionally the result in an exercise is used later.

The bibliography, although extensive, makes no claims to completeness; I have tried to refer mainly to the most recent works, and to works which themselves have good bibliographies. Especially, it should not be assumed that the number of papers listed under an author's name is any measure of the significance of his contributions to approximation theory. Thus, there are no entries for S. Bernstein, because his key works are referred to in TIMAN, which is in the bibliography (observe that a name written in capitals is a reference to the bibliography). Likewise, in connection with the brief discussion of rational approximation, I did not list the seminal papers of Newman, and Szüsz & Turán, but rather a later paper, FREUD_2, which in turn refers to these.

Chapters 2 and 3 contain a fairly systematic development of the theory of best uniform approximation, based (following RIVLIN & SHAPIRO) on elementary properties of convex sets. Thus, these chapters have a quite elementary character. The discussion of problems involving maximization of linear functionals side by side with problems of best approximation seems to me altogether natural, even inevitable, insofar as each class of problems is attacked by means of the same variational principles; moreover, the two types of problems are, in a certain sense, dual to one another.

Chapters 4 and 5 continue the discussion of best approximation, but for L^p norms. Here the treatment is less elementary, functional analysis (especially the Hahn-Banach theorem and the Riesz representation theorem) playing a central role. I do not aim at the full generality of e.g. SINGER_1, but stick to L^p spaces and concentrate on applications, largely in complex analysis, which is a fertile field for these methods. I chose mostly very simple applications, mainly to illustrate the methods, referring to the literature for applications having greater depth.

Chapter 6 contains standard material on Hilbert space, as it relates to approximation and extremal problems. I have stressed the usefulness of reproducing kernels, giving numerous examples. Chapter 7 can be studied as a topic in Fourier analysis; I have included it because of applications to approximation theory, following here the tradition of AHIEZER_1. Chapter 7 contains some material not available elsewhere, partly taken from the unpublished dissertation of LOGAN, partly based on my own studies. I think it presents a lively field for further research.

Chapter 8 has somewhat the character of an essay, outlining in broad features the theory of "degree of approximation" from the standpoint of the geometry of function spaces, ideas due largely to Kolmogorov and other Soviet mathematicians. This chapter more than the others needs supplementing from outside sources. Chapter 9 gives a rather general form of the theory of approximation by means of convolution integrals. It is likely to appear unmotivated to the unprepared reader (although, in fact, it developed out of the study of many special problems) and a preliminary reading of part of $SHAPIRO_1$ might be advisable.

The chapter by Jan Boman deals with saturation theory for convolution integrals, a topic closely related to Chapter 9. His treatment of this problem is very incisive, making maximal use of the powerful formalism of distribution theory. The chapter by Torbjörn Hedberg, based on a seminar talk of his, presents a new, and to my way of thinking very simple, proof (which also lends itself to generalizations) of Kolmogorov's famous theorem on superpositions of functions.

ACKNOWLEDGEMENTS. First and foremost I wish to thank Civ.ing. B.O. Björnestål of the Royal Institute of Technology, for his encouragement and assistance in the preparation of these notes. It was because of his initiative that the writing of the notes was undertaken. He wrote first drafts of several chapters and also compiled the bibliography. In addition, he gave selflessly of his time, scrutinizing the whole manuscript and correcting numerous errors. My thanks are also due to the faculty of the Mathematical Department at the Royal Institute of Technology, especially Prof. Bo Kjellberg, for the hospitality shown me, and for generously making available funds to pay for the typing of the manuscript.

Dr. Benjamin Logan, Jr., of the Bell Telephone Laboratories, kindly supplied me with a copy of his unpublished doctoral dissertation, on which part of Chapter 7 is based. In connection with Chapter 7, Yngve Domar and Carl Herz also contributed several valuable remarks. It is a pleasure to express my thanks to these people, and especially to Jan Boman and Torbjörn Hedberg for the excellent chapters they contributed to the present volume. Boman also generously made available to me an unpublished manuscript bearing on the material in Chapter 9, and I have also benefited very much from discussions with him.

These notes were written during July and August of 1970, in the beautiful surroundings of the Mittag-Leffler Institute. I am most grateful to Prof. Lennart Carleson for his kind invitation to me to live and work at that Institute. Last but not least, I wish to express my indebtedness to Fru Eva Rudh of Djursholms Ekeby for her painstaking and accurate typing of the manuscript.

Djursholm, August 1970.

NOTE. I will be grateful for any criticisms or comments received from readers of these notes, especially the correction of errors, or calling my attention to relevant works that I have ignored, for purposes of rectification in an eventual future edition.

CONTENTS

Chapter 1. Introduction

Concerning the nature of approximation theory, a good deal of "philosophy" is scattered throughout the following chapters (especially Chapter 8), so these introductory remarks shall be rather brief.

It is hard to say what "approximation theory" is, and perhaps there is no particular point in trying to define it. However, an author who has included the words "approximation theory" in the title of his book does find himself trying to decide whether this or that special topic qualifies for inclusion. "It's beautiful - but is it approximation theory?"

Certainly, if there is to be any body of theory worthy of being labelled "approximation theory", it should have some recognizable contours, and not merely be the union of all problems connected with approximation in one form or another. There is probably very little in common e.g. to Diophantine approximation and to the approximate solution of partial differential equations, or to either of these topics and the approximate location of the roots of a polynomial. There seems, however, to be general agreement that there is a recognizable, even if vaguely defined, subject called approximation theory; or what amounts to the same thing: among the myriad of approximation problems that confront the analyst, there are certain identifiable types that can be studied by common methods.

It seems that one can see two prototypical problems, around which much of the later work has crystallized. First, one has the Chebyshev problem of finding a polynomial of degree n which gives the best uniform approximation to a given continuous function on an interval. This is a problem of "best approximation", an extremal problem, and as such requires for its solution variational methods. From the standpoint of method, it is related to other sup norm extremal problems, e.g. maximize the derivative at a given point, among all polynomials of degree n bounded by one on a given interval. It is the choice of norm (here the sup norm) which gives its distinctive coloration to these problems, the methods employed and the results attained (e.g. the Chebyshev alternation theorem, cf. Chapter 2.) Experience has shown that the appropriate tool for sup norm extremal problems is the theory of convex sets. In like manner, various extremal problems for the L^2 norm are advantageously studied with the aid of the notions of orthogonality and orthogonal projection. (Of course, one should not interpret these remarks too rigidly, e.g. one could argue with equal justice that linear problems as a group show common features and employ common methods, as opposed to non-linear problems.) Experience also shows that extremal problems, even when the data is very explicit in character, can seldom be "solved" in an explicit form, and so the theory is largely devoted to questions of a general nature, e.g. existence, uniqueness, and characteristic properties of extremals. The main justifications for this theory, considering the implicit nature of the results, are (i) there is a set of extremal problems, of measure zero but very important for applications, which are solvable explicitly in the classical sense (these solutions are often characterized, moreover, by great elegance), (ii) the characterization of extremals usually is helpful in constructing "near extremals" or computational algorithms for the extremals, and (iii) the study of extremal problems, is a powerful tool in existence proofs, in accordance with some such scheme as: "The extremal problem P admits at least one solution (because of compactness); every extremal must have such and such properties (because of a variational argument); therefore, a function exists which has such and such properties". (Examples: Perron's solution of the Dirichlet problem; Carathéodory's proof of the Riemann mapping theorem.)

The second prototypical problem is that of Weierstrass, to show that every continuous function on a closed bounded interval can be uniformly approximated with arbitrarily small error by a polynomial. Here the Chebyshev approach has not thus far been very helpful, because the best approximating polynomial of a fixed degree is characterized too implicitly to permit any estimates. Here the best is the enemy of the good. This is a tendency that is visible throughout mathematical analysis: one learns that the solutions to problems can seldom be expressed explicitly in terms of familiar functions, and sets out to find estimates for them. This reflects itself in the present book in the polarity between the problems of "best approximation" and "good approximation". Historically, the second tendency came rather later; a long time elapsed before the Weierstrass theorem was put into a really quantitative form (by Jackson).

The Weierstrass theorem can be stated in the following form. Define for given f, continuous on I (where I = [- 1, 1], say),

$$(1) \qquad E_n(f) = \inf_{p \in \mathcal{P}_n} \|f - p\|$$

where $\| \ \|$ denotes sup norm, and $\mathcal{P}_n$ is the set of polynomials of degree at most n. Then

$$(2) \qquad \lim_{n \to \infty} E_n(f) = 0 \quad .$$

This suggests two kinds of generalizations: (i) find more general classes than $\mathcal{P}_n$ such that (2) continues to hold for the analogously defined $E_n(f)$, and (ii) try to say something more precise than (2). The first kind of generalization leads to the general problem of <u>completeness</u> (of course not necessarily restricted to the space C(I)) for a set of functions; the second to what could be called degree-of-approximation theory. Typical generalizations of the first kind are the theorems of Müntz (cf. Chapter 6) and of Stone; typical generalizations of the second kind are the theorems of Jackson (cf. Chapter 8). (In case (ii), no conclusion stronger than (2) is possible without additional assumptions (i.e. beyond continuity) about f.) Recently, as technique has developed, generalizations of the Weierstrass theorem have been proven which combine the features (i) and (ii) ("Müntz-Jackson theorems", cf. GANELIUS & WESTLUND).

What we have said thus far is applicable to the approximation of <u>functions</u>, or to extremal problems with a variable function. Another important area of approximation theory (although it finds little representation in the present book, outside of some results in Chapter 3) is concerned with the approximation of <u>functionals</u>, and extremal problems involving them. Here the prototypical problems are those of "quadrature" (i.e. of approximating to the integral of a (sufficiently nice) function by a linear combination of its values at selected points) and "interpolation" (i.e. of approximating the value of a function at a point from its values at certain other points). Again, these problems may be approached either in the spirit of "best approximation" or of "good approximation". Here, the latter tendency seems historically the older one (estimating the error in various concrete schemes for quadrature or interpolation), but there has been a lively interest in recent years in "optimal" functionals, "best" quadrature formulae, etc. Accounts of interpolation, differentiation and quadrature formulae from the older point of view are legion, cf. any work on numerical analysis. Concerning optimal functionals etc., there is some material in DAVIS, GOLOMB & WEINBERGER, SARD, STECHKIN$_1$, VORONOVSKAYA. It is to be expected that, as mathematics grows in complexity, the "unknowns" in our problems become more complex kinds of entities (e.g. functionals, or operators, rather than functions). In the Kolmogorov problem of "widths" (cf. Chapter 8) one sets up an extremal problem where the "variable" ranges over all n-dimensional subspaces of a given normed linear space. Needless to say, investigations of this kind move more and more away from classical analysis and are studied in the framework of functional analysis.

Chapter 2. Best Uniform Approximation

2.1 Notation and Preliminary Results. The real line shall be denoted throughout these notes by R, and the real Euclidean n-space by R^n. The closed interval [- 1, 1] is denoted by I, and the unit circle by T. Q will always denote a compact Hausdorff space (which, in applications, is usually a compact set in R^n). The class of real algebraic (trigonometric) polynomials of degree $\leq$ n is denoted by $\mathcal{P}_n$ ($\mathcal{T}_n$) and C(Q) denotes the linear space of continuous functions on Q, made into a Banach space with the usual norm

$$\|f\| = \max_{x \in Q} |f(x)| .$$

Approximation in this norm is called uniform (or Chebyshev) approximation (occasionally minimax approximation).

Now, given $f \in C(Q)$ and a subset P of C(Q), a basic problem is to seek an element $p^* \in P$ such that $\|f - p^*\| \leq \|f - p\|$ holds, for all $p \in P$. Such an element, if it exists, is called an element of best approximation (or simply, a best approximation, or closest element) to f from P. The number $\|f - p^*\|$ is called the distance from f to P; more generally we write (whether the inf is attained or not)

$$\text{dist}\,(f, P) = \inf_{p \in P} \|f - p\| .$$

(We shall use these terms also when C(Q) is replaced by other normed spaces.)

Does there always exist an element of best approximation? If it exists, is it unique, and, can we characterize it in some other way (which may lead to a solution in concrete situations)? Without further hypotheses about P there would be little hope to give any answers. As a rule, we shall consider the case when P is a closed subspace of C(Q), usually finite-dimensional. In this case, the question of existence at least is easily answered, and indeed, in the full generality of normed spaces, by the following theorem.

2.1.1 Theorem. Let P be a finite-dimensional subspace of a linear normed space X. Then, for each $f \in X$, there exists in P an element of best approximation to f.

Proof. Let p_0 denote any fixed element in P. Define

$$B = \left\{p \in P: \|f - p\| \leq \|f - p_0\|\right\} .$$

Then,

$$\inf_{p \in P} \|f - p\| = \inf_{p \in B} \|f - p\| .$$

But the closed ball B, being in a finite-dimensional space, is compact and hence the continuous function $p \to \|f - p\|$, attains its minimum at some point in B, which proves the theorem. ◇

Remark. The finite-dimensionality is essential to the truth of the theorem, as we shall see later by examples.

Now, back to C(Q). What properties does an extremal p^* have? Observe that for all $p \in P$

$$\|f - p^* + p\| \geq \|f - p^*\| ,$$

and hence, putting $f - p^* = e$ (the "error"), we have

(1) $$\|e + p\| \geq \|e\| \text{ , all } p \in P .$$

(1) expresses the basic variational condition that arises in all linear extremal problems. Generally speaking, in each normed space, the condition (1) (for the norm in question) turns out to have certain characteristic consequences, which are the key to the study of best approximation in that particular norm. In a Hilbert space, (1) expresses the condition that e is orthogonal to a subspace P. Therefore, we will speak of (1) as an orthogonality relation in other cases too.

Thus, if P is any subspace of a normed linear space X, we may say: a necessary and sufficient condition that $p^* \in P$ be a closest element to $f \in X$, is that $e = f - p^*$ is orthogonal to P. This relationship of (generalized) orthogonality will be written thus:

$$e \perp_X P ,$$

or simply $e \perp P$ when it is clear to which normed space we are referring. (Observe that generalized orthogonality lacks the nice properties of Hilbert space orthogonality. Thus, if e is orthogonal to the one-dimensional space spanned by f, the corresponding statement with the roles of e and f reversed is, in general, false. Moreover, $e \perp P_1$ and $e \perp P_2$ need not imply $e \perp P_3$ where P_3 is the space spanned by $P_1 \cup P_2$. In particular, orthogonality of e to a subspace P is not (as in the Hilbert space case) reducible to element-wise orthogonality.)

2.2 A General Characterization and Polynomial Approximation. In the uniform norm, we can obtain a fairly effective characterization of orthogonality (whereby P need not be finite-dimensional, but the criterion will assume a sharpened form in the latter case).

2.2.1 Lemma. Let Q be a compact Hausdorff space and let $f \in C(Q)$, f not identically zero. Define

$$E^+ = E^+(f) = \left\{x \in Q: f(x) = + \|f\|\right\}$$

$$E^- = E^-(f) = \left\{x \in Q: f(x) = - \|f\|\right\} .$$

The necessary and sufficient condition that f be orthogonal to the subspace P of C(Q) is that the relations

$$(1) \qquad \begin{cases} p(x) > 0 & x \in E^+ \\ p(x) < 0 & x \in E^- , \end{cases}$$

cannot hold for any $p \in P$.

Remark. Condition (1) can be written as

$$f(x)\ p(x) > 0 \quad , \qquad x \in E^+ \cup E^- .$$

The extension to complex-valued functions is proved in essentially the same way and reads: f is orthogonal to P if and only if there exists no $p \in P$ such that

$$\mathrm{Re}\ (\overline{f(x)}\ p(x)) > 0$$

holds on $E = \left\{x: |f(x)| = \|f\|\right\}$.

Proof (real case). (i) Assuming $p \in P$ satisfies (1), we shall show that $\|f - \epsilon p\| < \|f\|$ for some $\epsilon > 0$ (and hence f is not $\perp$ P). By hypothesis $f(x)p(x) > 0$ on the closed (and hence compact) set $E = E^+ \cup E^-$. Therefore, fp attains a positive minimum (say 2δ) on E, and there exists an open set G containing E, such that

$$f(x)\ p(x) > \delta > 0 \quad , \qquad x \in G \ .$$

In the complement G' of G, which is also compact, $|f| < \|f\|$, so there exists $\alpha > 0$, such that

$$|f(x)| \leq (1 - \alpha)\ \|f\| \ , \qquad x \in G' .$$

Thus, for $x \in G$,

$$(f(x) - \epsilon p(x))^2 \leq \|f\|^2 - 2\epsilon\delta + \epsilon^2\ \|p\|^2 < \|f\|^2$$

if $\epsilon > 0$ is chosen small enough, and for $x \in G'$,

$$|f(x) - \epsilon p(x)| \leq (1 - \alpha)\ \|f\| + \epsilon\ \|p\| \ .$$

Thus, for suitably small ϵ

$$|f(x) - \epsilon p(x)| < \|f\|$$

throughout Q.

(ii) If f is not orthogonal to P, there exists $p \in P$ such that $\|f - p\| < \|f\|$, i.e.

$$|f(x) - p(x)| < \|f\| \quad , \qquad x \in Q \ .$$

Clearly, such a polynomial has to be positive on E^+ and negative on E^-. ◇

At this point, we stop to deduce a classical theorem of Chebyshev; another proof will be given below, on the basis of more general ideas developed there.

2.2.2 Theorem. *If* f *(real)* $\in C(I)$ *and for some* $p \in \mathcal{P}_n$, $f(x) - p(x)$ *attains its maximum modulus at* $n + 2$ *points of* I *with alternating signs, then* p *is a best approximation to* f *from* $\mathcal{P}_n$. *Conversely, if* $f \notin \mathcal{P}_n$, *then for any best approximation* p *to* f *from* $\mathcal{P}_n$, $f - p$ *must attain its maximum modulus at some set of* $n + 2$ *points of* I *with alternating signs.*

Proof. In view of what has been said, it is enough to prove

2.2.2.1 Theorem. *If* e *(real)* $\in C(I)$ *is not identically zero, then* $e \perp \mathcal{P}_n$ *if and only if* $e(x)$ *attains its maximum modulus at* $n + 2$ *points of* I *with alternating signs.*

Proof. (i) Suppose first e is not orthogonal to $\mathcal{P}_n$. Then, by Lemma 2.2.1, there exists $p \in \mathcal{P}_n$ positive on $E^+(e)$ and negative on $E^-(e)$. The hypothesis about e, and the continuity of p, imply that p vanishes at $n + 1$ points, and hence identically, a contradiction.

(ii) We shall present this portion of the proof somewhat concisely; it is mainly of historical interest in view of the alternate proof to be given later. Indeed, the formal complications which arise in the present "naive" argument may serve to illustrate why the later, general approach is desirable.

Let x_1 denote the smallest element in $E = E(e)$. Without loss of generality, we may suppose $x_1 \in E^+$. Consider now

$$F_1 = \left\{x:\ x \in E^- \ \text{ and } \ x > x_1\right\} .$$

If F_1 is not empty, it has a minimal element x_2. Similarly, the set

$$F_2 = \left\{x:\ x \in E^+ \ \text{ and } \ x > x_2\right\}$$

is either empty, or it has a minimal element x_3. Proceeding in this way, either we construct a sequence $x_1, \dots x_{n+2}$ belonging to E^+ and E^- alternately, in which case the proof is complete, or, for some $r \leq n + 1$, F_r is empty. By virtue of Lemma 2.2.1 it is enough to show that in the latter case $p \in \mathcal{P}_n$ can be constructed such that $p(x)\ e(x) > 0,\ x \in E$.

To this end, define numbers λ_i $(i = 1, 2, \dots r - 1)$ such that

a) $\qquad x_i < \lambda_i < x_{i+1} \qquad i = 1, 2, \dots r - 1$

b) $\qquad [\lambda_i, x_{i+1}) \quad$ contains no point of E.

It is easy to check that the construction is possible, and that $p(x) = \pm (x - \lambda_1) \dots$ $\dots (x - \lambda_{r-1})$ satisfies the requirements. $\Diamond$

As an illustration of the previous theorem we have:

<u>2.2.3 Corollary (Chebyshev).</u> <u>For arbitrary real</u> a_i

$$\max_{|x|\leq 1} |x^n + a_{n-1} x^{n-1} + \dots + a_0| \geq 2^{1-n}$$

<u>and the value</u> 2^{1-n} <u>is the largest for which the assertion is true</u>. <u>Equivalently</u>,

$$\text{dist } (x^n, \mathcal{P}_{n-1}) = \min_{p \in \mathcal{P}_{n-1}} \|x^n - p\| = 2^{1-n}.$$

<u>Proof</u>. Let p_n denote a polynomial in $\mathcal{P}_n$ with leading coefficient 1 and having least maximum modulus on I. By Theorem 2.2.2, since $p_n \perp \mathcal{P}_{n-1}$, the extremal property of p_n

is equivalent to: p_n attains its maximum modulus with alternating signs at $n + 1$ points of I. $|p_n(x)|$ cannot attain its maximum at more than $n - 1$ interior points of I, because otherwise $p_n'(x)$ would have to vanish at these points and hence identically. So, two of the points must be the endpoints of I. If we put $\|p_n\| = M_n$, we get the following differential equation for p_n (because both sides have the same degree and the same zeroes):

$$M_n^2 - p_n(x)^2 = C_n(1 - x^2)\ p_n'(x)^2 .$$

Comparing the coefficients of x^{2n} we see that $C_n = n^{-2}$, and the solution of the equation is:

$$p_n(x) = M_n \cos (n \operatorname{arc} \cos x + c).$$

It is easily seen that the right side is a polynomial only if c is an integer multiple of π, and since the leading coefficient of p_n is positive we must have

(1) $$p_n(x) = M_n \cos (n \operatorname{arc} \cos x) .$$

It is easy to check that

(2) $$\cos (n \operatorname{arc} \cos x) = ((x + \sqrt{x^2 - 1})^n + (x - \sqrt{x^2 - 1})^n)/2 .$$

In this expression the odd powers of $\sqrt{x^2 - 1}$ cancel, putting in evidence that $\cos (n \operatorname{arc} \cos x)$ is indeed an algebraic polynomial of degree n. From (1), (2) it is easy to see that the leading coefficient of p_n equals $2^{n-1} M_n$ (cf. the following paragraph), hence $M_n = 2^{1-n}$. ◇

2.2.3.1 The Chebyshev polynomials are defined by

(1 a) $$C_n(x) = \cos (n \operatorname{arc} \cos x)$$

or equivalently,

(1 b) $$C_n(\cos t) = \cos nt ,$$

$n \geq 0$, and are easily seen to satisfy the recurrence formula

$$C_{n+1}(x) = 2x\, C_n(x) - C_{n-1}(x) , \quad n \geq 1 ,$$

$$C_0(x) = 1 , \quad C_1(x) = x .$$

One sees inductively that the leading coefficient in $C_n(x)$ is 2^{n-1}, $n \geq 1$. The Chebyshev polynomials, which occur in many branches of applied mathematics, also satisfy many other functional equations and have other extremal properties (cf. Chapter 3).

Remarks. A shorter, but less well motivated, proof of 2.2.3 could have been given by just verifying at once that $C_n(x)$ (as defined, say, by (1 b)) is orthogonal to $\mathcal{P}_{n-1}$, with the aid of Theorem 2.2.2.1. Here we were able, by an *ad hoc* argument, to deduce a differential equation satisfied by the extremal polynomial, and hence deduce the latter; but in most cases where problems of best approximation have been solved explicitly, the procedure is to "guess" an extremal function, based on properties which extremals are known in general to have, and then verify that it satisfies conditions known to be sufficient for an extremal. This theme will recur often in these lectures.

It is worth emphasizing that (as the previous analysis shows) the Chebyshev polynomial C_n is characterized (uniquely, apart from a constant factor) among all non-constant elements of $\mathcal{P}_n$ by the property that it attains its maximum modulus at the greatest possible number $(n + 1)$ of points of I.

We now consider the question of uniqueness of the best approximation in $\mathcal{P}_n$.

2.2.4 Theorem. *For each real* f *in* C(I) *the best approximation to* f *from* $\mathcal{P}_n$ *is unique*.

The proof follows immediately from the following lemma.

2.2.5 Lemma. *Let* $p \in \mathcal{P}_n$ *and let the* $n + 2$ *points* x_i *satisfy* $-1 \leq x_1 < x_2 < \ldots < x_{n+2} \leq 1$. *If* $(-1)^{i-1}\, p(x_i) \geq 0$, $i = 1, \ldots, n + 2$, *then* $p \equiv 0$. Indeed, assume the lemma holds; since the theorem is clearly true when $f \in \mathcal{P}_n$, we may suppose $f \notin \mathcal{P}_n$. Let p_1 be a best approximation to f from $\mathcal{P}_n$. Then $f - p_1$ attains its maximum modulus at $n + 2$ points x_i with successive changes of sign. Without loss of generality, we may suppose $f(x_1) - p_1(x_1) > 0$. If now, for some $p_2 \in \mathcal{P}_n$,

$$\|f - p_2\| = \|f - p_1\| , \quad \text{i.e.}$$

$$\|(f - p_1) - (p_2 - p_1)\| = \|f - p_1\|$$

then clearly $(p_2 - p_1)(x_1) \geq 0$, $(p_2 - p_1)(x_2) \leq 0$, etc., hence by the lemma $p_2 - p_1 \equiv 0$. ◇

For the proof of Lemma 2.2.5 we need the following useful lemma.

2.2.6 Lemma. Given n + 2 distinct points $\{x_i\}$ of I satisfying $x_i < x_{i+1}$, there exist numbers $c_i > 0$, such that

$$\sum_{i=1}^{n+2} (-1)^{i-1} c_i\, p(x_i) = 0$$

for all $p \in \mathcal{P}_n$.

Proof. On $\mathcal{P}_n$, let us define linear functionals

$$L_i(p) = p(x_i) \quad , \quad i = 1, \ldots, n+2 .$$

Since the dimension of $\mathcal{P}_n$ is n + 1, the n + 2 functionals L_i cannot be linearly independent, i.e. there exist b_i, i = 1, ..., n + 2, not all zero, such that

$$\sum_{i=1}^{n+2} b_i\, p(x_i) = 0 \qquad p \in \mathcal{P}_n .$$

Now, setting

$$p(x) = (x - x_1) \ldots (x - x_{K-1})(x - x_{K+2}) \ldots (x - x_{n+2})$$

we have

$$\text{(1)} \qquad \sum_{i=1}^{n+2} b_i\, p(x_i) = b_K\, p(x_K) + b_{K+1}\, p(x_{K+1}) = 0 .$$

But as this particular polynomial does not change sign between x_{K-1} and x_{K+2}, it follows from (1) that the b_i alternate in sign. Therefore $c_i = (-1)^{i-1} b_i$ satisfies the requirements. ◇

Proof of Lemma 2.2.5. If p satisfies the hypotheses of the lemma, then, according to Lemma 2.2.6, there exist strictly positive numbers c_i such that $\sum_{i=1}^{n+2}(-1)^{i-1}c_i p(x_i) = 0$. As we have a zero sum with non-negative terms, each term must be zero, i.e. $p(x_1) = p(x_2) = \ldots = p(x_{n+2}) = 0$, which is more than is required for the conclusion that p = 0. ◇

2.2.7 An alternate proof. A second proof of Theorem 2.2.4 may be given, which embodies a useful method. Let $p, q \in \mathcal{P}_n$ be polynomials of best approximation to $f \in C(I)$. Define $r(x) = (p(x) + q(x))/2$; then $r \in \mathcal{P}_n$ and

$$\|f - r\| \leq \frac{1}{2} \|f - p\| + \frac{1}{2} \|f - q\| = d ,$$

where $d = \text{dist}(f, \mathcal{P}_n)$. Hence $\|f - r\| = d$, and r is a best approximation to f. Now, consider a point ξ where

$$f(\xi) - r(\xi) = + d .$$

Since

$$d = \frac{1}{2}(f(\xi) - p(\xi)) + \frac{1}{2}(f(\xi) - q(\xi)) \leq (d/2) + (d/2) = d ,$$

we must have

$$f(\xi) - p(\xi) = f(\xi) - q(\xi) = d .$$

Thus, $p(\xi) = q(\xi)$, and $p(x) - q(x)$ vanishes at any point where $f(x) - r(x) = + d$. This is of course also true where $f(x) - r(x)$ equals $- d$. But according to Theorem 2.2.2 $|f(x) - r(x)|$ attains its maximum at more than $n + 1$ points, and since $p - q$ vanishes at those points, it vanishes identically. ◇

2.3 The Main Theorem. A set E in a linear space is, we recall, convex if $x, y \in E$ and $0 \leq c \leq 1$ implies $cx + (1 - c)y \in E$. The convex hull of a set A, hull (A), is the intersection of all convex sets containing A. Equivalently: hull $(A) = \left\{\sum c_i x_i\right\}$ where $\left\{x_i\right\}$ is a finite subset of A, $c_i > 0$ and $\sum c_i = 1$. Linear combinations of this kind can be called convex linear combinations.

We assume known the most elementary properties of convex sets, notably the concept of hyperplane, and the principle of the separating hyperplane. The following lemma is slightly more special and we give a proof.

2.3.1 Lemma (Carathéodory). *If* $A \subset R^n$, *then every point of* hull (A) *can be written as a convex linear combination of at most* $n + 1$ *points of* A.

Proof. Let $x \in A$. After a translation, if necessary, we may assume that $x = 0$. Suppose

(1)
$$0 = \sum_{i=1}^{r} \alpha_i x_i \qquad x_i \in A, \quad \alpha_i > 0, \quad r > n + 1 .$$

Since $r > n + 1$, the elements $x_2, x_3, \ldots, x_r$ are linearly dependent, i.e. there exist β_i, $i = 2, \ldots, r$, not all zero, such that

$$\sum_{i=2}^{r} \beta_i x_i = 0 .$$

Define β_1 to be 0. For all real λ we have

$$0 = \sum_{i=1}^{r} \alpha_i x_i + \lambda \sum_{i=1}^{r} \beta_i x_i = \sum_{i=1}^{r} (\alpha_i + \lambda \beta_i) x_i .$$

When $\lambda = 0$, all terms in the last sum are positive. Now, define $c = \min |\alpha_i/\beta_i|$, the min being over those i with $\beta_i \neq 0$. Choosing a value of j such that $|\alpha_j/\beta_j| = c$, and setting $\lambda = -\alpha_j/\beta_j$, at least one of the numbers $\alpha_i + \lambda\beta_i$ is zero, and all are non-negative. Also $\alpha_1 + \lambda\beta_1 = \alpha_1 > 0$. We have thus obtained a representation of the same kind as (1) but with $s \leqslant r - 1$ terms. If $s > n + 1$ the process can be repeated, and after a finite number of steps we obtain the representation postulated in the lemma. $\diamond$

__Remark.__ If x belongs to the boundary of hull (A) then n points suffice (exercise).

Let P be an n-dimensional subspace of real-valued (complex-valued) C(Q), and as before let $E = E(f)$ denote $\{x \in Q: |f(x)| = \|f\|\}$. Note that when C(Q) is complex, P is isomorphic to a 2n-dimensional real space.

__2.3.2 Theorem (Main Theorem).__ __Necessary and sufficient that the function__ $f \in C(Q)$, __not identically zero, be orthogonal to__ P, __is that there exist points__ $x_1, \ldots, x_r$ __in__ E, __where__ $1 \leqslant r \leqslant n + 1$ __when__ C(Q) __is real, and__ $1 \leqslant r \leqslant 2n + 1$ __when__ C(Q) __is complex, and positive numbers__ $c_1, \ldots, c_r$, __such that__

$$\sum_{i=1}^{r} c_i \overline{f(x_i)}\, p(x_i) = 0 , \qquad p \in P . \tag{1}$$

__Proof.__ (i) __Sufficiency.__ Suppose (1) holds and that $\sum_{i=1}^{r} c_i = 1$. As $|f(x_i)| = \|f\|$, we have

$$\|f\|^2 = \sum_{i=1}^{r} c_i \overline{f(x_i)} \cdot f(x_i) = \sum_{i=1}^{r} c_i \overline{f(x_i)} (f(x_i) - p(x_i)) \leqslant$$

$$\leqslant \|f\| \cdot \sum_{i=1}^{r} c_i \cdot \max_j |f(x_j) - p(x_j)| \leqslant \|f\| \cdot \|f - p\|$$

for all $p \in P$, i.e. $f \perp P$.

(<u>Remark</u>. Observe that the condition that $r \leq n+1$ (resp. $2n+1$) was not needed for this part of the proof, i.e. <u>the sufficiency of (1) is valid with no hypothesis about</u> r.)

(ii) <u>Necessity</u>. Suppose $f \perp P$. Let $\varphi_1, \ldots, \varphi_n$ be a basis for P and consider the map $T: Q \to R^n$ (resp. $\mathbb{C}^n$) defined by

$$T(x) = \overline{f(x)}\,(\varphi_1(x), \ldots, \varphi_n(x)).$$

First we claim that the origin lies in the convex hull of TE, the image of E under T. Indeed, suppose on the contrary, there were to exist a hyperplane separating TE and 0, i.e. complex numbers $a_1, \ldots, a_n$ such that

$$\text{(2)} \qquad \operatorname{Re}\left\{\sum_{i=1}^{n} a_i\, \overline{f(x)}\, \varphi_i(x)\right\} > 0, \qquad x \in E.$$

Writing $p_0(x) = \sum_{i=1}^{n} a_i\, \varphi_i(x)$, (2) and Lemma 2.2.1 (and the remark which followed it) imply that f is not orthogonal to P, a contradiction.

Hence the origin is in hull (TE), and by Carathéodory's lemma there exist points $x_1, \ldots, x_r$ in E, where $1 \leq r \leq n+1$ (resp. $2n+1$), and $c_i > 0$ such that

$$\sum_{i=1}^{r} c_i\, \overline{f(x_i)}\, \varphi_\nu(x_i) = 0 \qquad \nu = 1, \ldots, n.$$

i.e.

$$\sum_{i=1}^{r} c_i\, \overline{f(x_i)}\, p(x_i) = 0 \qquad p \in P,$$

as was to be proved. ◇

<u>2.3.3 Corollary</u>. <u>If</u> $f \perp P$ <u>in</u> C(Q) <u>then</u> $f|_F \perp P|_F$ <u>in</u> C(F) <u>where</u> F <u>is some subset of</u> Q, <u>consisting of at most</u> n + 1 (<u>resp</u>. 2n + 1) <u>points</u>. (As usual, $f|_F$ denotes the restriction of f to F.)

<u>Proof</u>. Simply take F to be the set $\{x_1, \ldots, x_r\}$ constructed in Theorem 2.3.2. ◇

A slightly different formulation of the corollary is the following:

<u>2.3.3.1 Theorem</u>. (<u>Generalized de la Vallée Poussin theorem</u>). <u>If, under the hypotheses of Theorem 2.3.2</u>, p^* ($\in P$) <u>is a best uniform approximation to</u> $f \in C(Q)$, <u>it</u>

remains a best uniform approximation to f when all functions in question are restricted to a suitable subset of Q having at most n + 1 (resp. 2n + 1) points.

2.3.3.2 Reformulation in terms of integrals. Sums like $\sum_{i=1}^{r} c_i\, p(x_i)$ can conveniently be rewritten as $\int p(x)\, d\rho(x)$, where ρ is the discrete (or "atomic") measure composed of the point-masses ("atoms") c_i at x_i. For example, the Main Theorem can be reformulated as follows:

$f \in C(Q)$ is orthogonal to P, an n-dimensional subspace of C(Q), if and only if there exists a non-null positive measure ρ whose support consists of r points of E(f), $1 \leq r \leq n + 1$ $(1 \leq r \leq 2n + 1)$, such that $\overline{f(x)}\, d\rho(x)$ "annihilates" P, i.e.

$$(1) \qquad \int p(x)\, \overline{f(x)}\, d\rho(x) = 0 \qquad p \in P .$$

(Without loss of generality we can of course assume $\int d\rho = 1$.).

This reformulation is not only a notational convenience but, as we shall see, is essential in generalizations where P is no longer finite-dimensional. Moreover, (1) with any non-null positive measure (not necessarily discrete) is sufficient for $f \perp P$; this remark is often useful, even when P is finite-dimensional.

2.3.4 Theorem. Let Q be a compact Hausdorff space, and P a separable subspace of C(Q). Let $f \in C(Q)$ be given, $f \notin P$, and let $E = \{x \in Q\colon |f(x)| = \|f\|\}$. Necessary and sufficient that f be orthogonal to P is that there exists a positive measure ρ supported by some closed subset F of E such that

$$(1) \qquad \int p(x)\, \overline{f(x)}\, d\rho(x) = 0 \qquad p \in P .$$

Proof. (i) Necessity. Since P is separable there exists a sequence of finite-dimensional subspaces R_n of P, such that the dimension of R_n is n and $\bigcup_{n=1}^{\infty} R_n$ is dense in P. For each n there exists a positive measure ρ_n of total mass one supported by some finite subset (depending on n) of E such that $\int p\, \overline{f}\, d\rho_n = 0$ for $p \in R_n$. By a standard compactness theorem, there exists a positive measure ρ, supported in E, and a subsequence of the ρ_n (which, for notational convenience, we continue to denote by ρ_n) that converges to ρ in the so-called weak*-topology of the space of measures (dual space of C(Q)), hence

$$\lim_{n \to \infty} \int p\,\overline{f}\,d\rho_n = \int p\,\overline{f}\,d\rho$$

for all $p \in P$. Now since $\bigcup_{n=1}^{\infty} R_n$ is (uniformly) dense in P and for $q \in \bigcup_{n=1}^{\infty} R_n$ we have

$$\lim_{n \to \infty} \int q\,\overline{f}\,d\rho_n = 0 ,$$

(1) holds, and the first part of the theorem is proved.

(ii) <u>Sufficiency</u>. If ρ ($\int d\rho = 1$) satisfies the hypotheses of the theorem, we have the standard computation (here P can be any closed subspace):

$$\|f\|^2 = \int \overline{f} \cdot f d\rho = \int \overline{f}(f - p)\, d\rho \leq \|f\|\ \|f - p\| . \qquad \diamond$$

<u>2.3.5 Remark</u>. The hypothesis of <u>separability</u> of P in the preceding theorem can be eliminated, if we are willing to borrow rather more from functional analysis. Indeed, if $f \perp P$, i.e. dist $(f, P) = \|f\|$, then by a standard corollary of the Hahn-Banach theorem (cf. Chapter 4, where the same argument is given for L^p spaces), there is a linear functional on C(Q) which vanishes on P, has norm $\|f\|^{-1}$ and takes the value 1 on f; hence by the Riesz representation theorem there is a complex measure μ on Q such that $\int p\, d\mu = 0$, $p \in P$, and

$$1 = \int f\, d\mu = \left|\int f d\mu\right| \leq \|f\| \int |d\mu| = 1$$

and the conditions for equality here imply that $d\mu = \overline{f}\, d\rho$ where ρ is a positive measure supported on $\{x : f(x) = \|f\|\}$.

We have preferred, for methodological simplicity, to develop the theory on an elementary basis (following RIVLIN & SHAPIRO), since <u>for applications to finite-dimensional spaces</u> that is perfectly adequate. For an alternate development based on the Hahn-Banach theorem and Riesz representation theorem, cf. ZUHOVITSKI. The method followed here yields as by-products <u>special cases of</u> the theorems of Hahn-Banach and Riesz. Cf. also HINTZMAN.

<u>2.3.6 Exercises</u>. a) Let Q = unit circle. Show that a function invariant with

respect to rotation through $2\pi/n$ is orthogonal to the subspace of trigonometric polynomials of degree at most $n - 1$. Cf. 4.2.4.

b) Let $Q = [-1, 1]$. Prove that an even function is orthogonal to the (subspace consisting of) odd functions, and an odd function is orthogonal to the even functions.

c) Let Q = unit sphere in R^n. Prove that the function $f(x) = x_1 x_2 \dots x_n$ is orthogonal in $C(Q)$ to the subspace of polynomials of degree at most $n - 1$ in the n variables.

2.3.7 Theorem. (Duality Theorem). Let P be an n-dimensional subspace of the real (complex) space $C(Q)$. Then, for all $f \in C(Q)$,

$$(1) \qquad \min_{p \in P} \|f - p\| = \max_{\mu} \frac{|\int f \, d\mu|}{\int |d\mu|},$$

where μ ranges over all discrete measures with $r \leq n + 1$ (resp. $2n + 1$) points of support, annihilating P.

Remark. This theorem, except for the restriction on the support is a simple consequence of the Hahn-Banach theorem, combined with F. Riesz' theorem on the representation of linear functionals on $C(Q)$; without the restriction on the support it is valid without assuming P finite-dimensional (and min replaced by inf). See Chapter 5.

Proof of Theorem 2.3.7. (i) If $\int p \, d\mu = 0$, all $p \in P$, and p^* is a best approximation to f (which we know exists), then

$$|\int f \, d\mu| = |\int (f - p^*) d\mu| \leq \|f - p^*\| \int |d\mu| ,$$

i.e.

$$\min_{p \in P} \|f - p\| \geq \frac{|\int f \, d\mu|}{\int |d\mu|}.$$

which implies half of (1).

(ii) Since $f - p^* \perp P$, there exists, according to Theorem 2.3.2, a positive measure ρ, with support at $1 \leq r \leq n + 1$ (resp. $2n + 1$) points of $E(f - p^*)$ such that

$$\int \overline{(f(x) - p^*(x))}\; p(x)\; d\rho(x) = 0 \qquad p \in P.$$

Taking $d\mu_o(x) = \overline{(f(x) - p^*(x))}\; d\rho(x)$, we have

$$\left|\int f\, d\mu_o\right| = \left|\int f\; \overline{(f - p^*)}\; d\rho\right| = \left|\int (f - p^*)\; \overline{(f - p^*)}\; d\rho\right| = \|f - p^*\|^2 \int \cdot\; d\rho$$

If we divide this expression by $\int |d\mu_o| = \|f - p^*\| \int d\rho$, we obtain

$$\|f - p^*\| = \frac{\left|\int f\, d\mu_o\right|}{\int |d\mu_o|}$$

and, since μ_o annihilates P, the other half of (1) is proved. ◇

By means of the Duality Theorem one can give an explicit formula for the distance from f in C(I) to certain finite-dimensional subspaces, e.g. algebraic polynomials of degree n; see Theorem 2.5.1 (and Corollary 2.5.2).

<u>2.4 Haar Spaces and Unicity Theorems</u>. From Theorem 2.2.4 we know that the best approximation from $\mathcal{P}_n$ to $f \in C(I)$ is unique. As will be seen, the unicity still holds when I is replaced by any compact subset of R. However, we do not in general have unicity if $\mathcal{P}_n$ is replaced by a more general set of functions, e.g. n merely linearly independent functions. This may be seen from the simplest examples; a very general kind of counter-example will be given below, in connection with Haar's theorem. We begin with some generalities.

Let Q be a compact Hausdorff space (with at least n points) and let P be the subspace of C(Q) spanned by the functions $g_1, \ldots, g_n$. Linear combinations of the g_i's are called <u>generalized polynomials</u>. Further, it is assumed that the functions g_i satisfy a rather strong independence condition, the so called <u>Haar condition</u> (such a set of functions is also sometimes termed a <u>Chebyshev system</u>). The continuous functions $g_1, \ldots, g_n$ are said to satisfy the Haar condition (or, P is said to be a <u>Haar</u> <u>n-space</u>) <u>if every generalized polynomial, not identically zero, has at most</u> n - 1 <u>zeroes</u>.

Equivalent formulations. The set of continuous functions $\{g_1, \ldots, g_n\}$ satisfies the Haar condition if and only if any of the following conditions I, II, III holds:

I.

$$D(x_1, \ldots, x_n) = \begin{vmatrix} g_1(x_1) & \ldots & g_1(x_n) \\ \vdots & & \vdots \\ g_n(x_1) & \ldots & g_n(x_n) \end{vmatrix}$$

(generalized Vandermonde determinant) is non-zero for any set of n distinct points $x_1, \ldots, x_n \in Q$.

II. There exists no measure annihilating P whose support consists of less than n + 1 points.

III. For all sets of distinct points $x_1, \ldots, x_n$ and complex numbers $c_1, \ldots, c_n$, the interpolation problem

$$p(x_i) = c_i \qquad i = 1, \ldots, n ,$$

is uniquely solvable in P.

Exercise. Prove that these statements are equivalent to the original definition.

For real-valued differentiable functions on I we have also the following characterization: Let $g_1, \ldots, g_n$ have continuous derivatives of order n on I. Then, the Haar condition is satisfied if and only if the Wronski determinant

$$W(g_1, \ldots, g_n)(t) = \begin{vmatrix} g_1(t) & \ldots & g_1^{(n)}(t) \\ \vdots & & \vdots \\ g_n(t) & \ldots & g_n^{(n)}(t) \end{vmatrix}$$

is non-vanishing on I.

For a proof, see the book KARLIN & STUDDEN pp. 376 ff. (this book gives an extensive and up-to-date treatment of Haar spaces).

The following is a basic property of real Haar spaces on I.

2.4.1 Lemma. Let P be a real Haar n-space on I and let n - 1 points x_i, $-1 < x_1 < \ldots < x_{n-1} < 1$, be given. Then there exists a non-zero element in P that vanishes at those points and changes sign at each point.

Remark. In other words, the (essentially unique) polynomial with the n - 1 prescribed zeroes changes sign at each zero.

Proof. Let $g_1, \ldots, g_n$ be a basis for P and consider the determinant (generalized polynomial)

$$D(x_1, \ldots, x_{n-1}, x) = \begin{vmatrix} g_1(x_1) & \cdots & g_1(x_{n-1}) & g_1(x) \\ \vdots & & \vdots & \vdots \\ g_n(x_1) & \cdots & g_n(x_{n-1}) & g_n(x) \end{vmatrix} .$$

Clearly D has the required zeroes. Now, select a value of k, where $1 \leq k \leq n - 1$. We must show that $D(x_1, \ldots, x_{n-1}, x)$ changes sign at $x = x_k$ or, what is the same thing, that $D(x_1, \ldots, x_{n-1}, x_k + y)$ changes sign at $y = 0$. Now, for small $|y|$, this determinant has the same sign as

$$D(x_1, \ldots, x_{k-1}, x_k - y, x_{k+1}, \ldots, x_{n-1}, x_k + y)$$

because the latter arises from the former by changing continuously the entries in the kth column in such a way that the determinant always has a real non-zero value. The last expression is, however, an odd function of y, which proves the lemma. ◇

Lemma 2.4.1 also holds on every set of n - 1 distinct points on the circle (same proof). As a consequence, we have: On the circle there exist no real Haar spaces of even dimension.

2.4.2 Examples. a) The (n + 1)-dimensional spaces $\mathcal{P}_n$ form Haar n + 1-spaces on every compact $Q \subset R$. $\mathcal{P}_n$ is also a Haar (n + 1)-space when considered as a space of complex-valued functions on an arbitrary compact subset of the plane.

b) The space $\mathcal{T}_n$ of trigonometrical polynomials of degree not exceeding n forms a Haar (2n + 1)-space on the "circle group", i.e. the real numbers (mod 2π). This is so because every element in $\mathcal{T}_n$ can be written

$$\sum_{-n}^{n} c_k e^{ikx} = z^{-n} p_{2n}(z) , \qquad z = e^{ix} ,$$

and the ("algebraic", complex) polynomial $p_{2n}(z)$, of degree 2n (or lower), cannot have more than 2n zeroes without being identically zero. It is also possible to identify real $\mathcal{T}_n$ as the set of restrictions to the circle $x^2 + y^2 = 1$ of real algebraic polynomials in x, y of degree not exceeding n (i.e. of linear combinations of monomials $x^i y^j$ with $0 \leq i + j \leq n$).

__Exercise.__ Verify this, and also that the analogous space of restrictions to the ellipse $(x/a)^2 + (y/b)^2 = 1$ is a Haar 2n + 1-space. (Hint: use Bézout's Theorem).

__2.4.3 Theorem.__ (__Haar's uniqueness theorem__). __Let__ P __be a finite-dimensional subspace of__ C(Q), __the linear space of continuous functions on the compact Hausdorff space__ Q. __For each__ $f \in C(Q)$, __there is a unique best approximation to__ f __from__ P, __if and only if__ P __is a Haar space.__

__Proof.__ (i) Suppose P is a Haar n-space and that p_1 and p_2 are best approximations to some $f \in C(Q)$. Then also $p_3 = (p_1 + p_2)/2$ is a best approximation. But whenever

$$f(x) - p_3(x) = \pm \|f - p_3\| , \tag{1}$$

then $p_1(x) = p_2(x) = p_3(x)$, so it suffices to show that (1) holds at n points. As $f - p_3 \perp P$, we know from Theorem 2.3.2 that $f(x) - p_3(x)$ attains its maximum modulus at r points that support an annihilating measure for P. Consequently, $r \geq n + 1$.
(ii) If P is not a Haar space there exists a set of n points $\{x_1, \ldots, x_n\}$ such that the system of homogenous linear equations

$$\begin{pmatrix} g_1(x_1) & \cdots & g_1(x_n) \\ \vdots & & \vdots \\ g_n(x_1) & \cdots & g_n(x_n) \end{pmatrix} \begin{pmatrix} a_1 \\ \vdots \\ a_n \end{pmatrix} = 0 , \tag{2}$$

where $\{g_1, \ldots, g_n\}$ is a basis for P, has a non-zero solution. Then also the homogenous system formed with the transposed matrix has a non-zero solution, i.e. there exist constants b_i, not all zero, such that $\sum_{i=1}^{n} b_i \, g_i(x_j) = 0$, $j = 1, \ldots, n$. Thus, if $q = \sum_{i=1}^{n} b_i \, g_i$, we have $q(x_j) = 0$, $j = 1, \ldots, n$. Normalize q to have norm one. Now let $f \in C(Q)$ be such that $\|f\| = 1$ and $f(x_j) = \overline{\operatorname{sgn} a_j}$, $j = 1, \ldots, n$ (such an f can always be constructed on a compact Hausdorff space). Setting $h(x) = f(x)(1 - |q(x)|)$, we have $h(x_j) = \overline{\operatorname{sgn} a_j}$, $j = 1, \ldots, n$. We claim that if $p \in P$, then $\|h - p\| \geq 1$. Indeed, if $\|h - p\| < 1$, then $\operatorname{sgn} p(x_i) = \operatorname{sgn} h(x_i) = \overline{\operatorname{sgn} a_i}$, which contradicts (2), i.e. the fact that the measure with mass a_i at x_i annihilates P. Finally, for all λ, $0 \leqslant \lambda \leqslant 1$, λq is a best approximation to h:

$$|h(x) - \lambda q(x)| \leqslant |f(x)|(1 - |q(x)|) + \lambda|q(x)| \leqslant 1 + (\lambda - 1)|q(x)| \leqslant 1$$

for all x, i.e. $\|h - \lambda q\| = \text{dist}(h, P)$. Thus the b.a. is not unique. ◇

In connection with the previous theorem it is natural to ask which topological spaces Q possess non-trivial Haar spaces. It was already noted by HAAR that a compact subset Q of R^n, $n \geqslant 2$ cannot contain interior points if it carries a non-trivial <u>real</u> Haar 2-space. Because otherwise, one could, by a continuous displacement <u>within</u> Q interchange two points (x_1, x_2) without "collision"; this interchanges the columns (and hence changes the sign) of the determinant

$$\begin{vmatrix} g_1(x_1) & g_1(x_2) \\ g_2(x_1) & g_2(x_2) \end{vmatrix}$$

while allowing the entries to vary continuously, hence the determinant must vanish somewhere "in between", contradicting the Haar condition. Clearly the same argument applies if Q contains three line segments which meet at a common vertex (or a homeomorph of this configuration) since we can make the desired interchange of two points lying on different segments by the familiar procedure of "shunting railroad cars".

The complete characterization of compact Hausdorff spaces that possess non-trivial <u>real</u> Haar spaces is due to Mairhuber (see CURTIS) and is as follows: <u>Necessary and sufficient for a compact Hausdorff space</u> Q <u>to possess non-trivial real Haar n-spaces</u> P, <u>for</u> $n \geqslant 2$, <u>is that</u> Q <u>be homeomorphic to a subset of a circle</u>. <u>If the dimension of</u> P <u>is even, the subset must be proper</u>. For a study of the analogous problem for complex Haar spaces, see SCHOENBERG & YANG.

The Haar n-spaces are characterized by the fact that there exists no annihilating measure whose support contains less than n + 1 points. There exist of course annihilating measures supported by n + 1 points, and Lemma 2.2.6 says that in the special case of the Haar n-spaces $\mathcal{P}_{n-1}$ the masses of these measures alternate in sign. In fact, this is a feature common to <u>all</u> real Haar spaces on I (or on any interval, or the circle).

2.4.4 Theorem. Let P be a real Haar n-space on a compact Hausdorff space Q. Any set of n + 1 distinct points in Q is the support of some non-zero P-annihilating measure, which is unique up to a constant factor. In case Q = I, the signs of the "masses" of this measure are opposite at each pair of adjacent points.

Proof. What is to be shown is that the system

$$\begin{pmatrix} g_1(x_1) & \cdots & g_1(x_{n+1}) \\ \vdots & & \vdots \\ g_n(x_1) & \cdots & g_n(x_{n+1}) \end{pmatrix} \begin{pmatrix} a_1 \\ \vdots \\ a_{n+1} \end{pmatrix} = 0 ,$$

where $\{g_1, \ldots, g_n\}$ is a basis for P, has a non-null solution, unique up to multiplication by a constant, and that if $-1 \leqslant x_1 < \ldots < x_{n+1} \leqslant 1$ (in the case Q = I), then $\operatorname{sgn} a_{i+1} = -\operatorname{sgn} a_i$, $i = 1, \ldots, n$. Since the Haar condition obtains, all but the last clause is contained in the earlier discussion. So, taking now Q = I, we consider non-zero a_i such that $\sum_{i=1}^{n+1} a_i\, p(x_i) = 0$ for all generalized polynomials $p \in P$. Choosing p to be a non-trivial element of P with the zeroes $x_1, \ldots, x_{k-1}, x_{k+2}, \ldots, x_{n+1}$ and (necessarily) no others in I, we have $a_k\, p(x_k) + a_{k+1}\, p(x_{k+1}) = 0$, and since $p(x_k)$ and $p(x_{k+1})$ have the same sign it follows that $\operatorname{sgn} a_{k+1} = -\operatorname{sgn} a_k$. ◇

In our next theorem we show that orthogonality to a Haar subspace has very strong consequences, reminiscent of the situation in Hilbert spaces. An analogous result for nonlinear approximating families is in BARRAR & LOEB.

2.4.5 Theorem. Let Q be a compact Hausdorff space, and P a Haar subspace of C(Q). Let $f \in C(Q)$ satisfy $f \perp P$. Then, for every $q \in P$ we have

(i) $$\|f + q\| \geqslant (\|f\|^2 + a\,\|q\|^2)^{1/2} .$$

Moreover, in the real-valued case we have

(ii) $$\|f + q\| \geqslant \|f\| + b\,\|q\| .$$

Here a, b denote constants depending on f and P, but not on q.

Proof. Let $A = \|f + q\| - \|f\|$, so $A \geqslant 0$. Now, by Theorem 2.3.2, f attains its maximum modulus M at points $x_1, \ldots, x_r$ and there exist positive numbers c_i such that for every $p \in P$

$$(1) \qquad \sum_{i=1}^{r} c_i \, \overline{\sigma}_i \, p(x_i) = 0$$

where $\sigma_i = \operatorname{sgn} f(x_i)$. Since $\|f + q\| = \|M\| + A$, we have

$$|M\sigma_i + q(x_i)| \leqslant M + A \qquad i = 1, \ldots, r ;$$

hence

$$(2) \qquad 2M \operatorname{Re}(\overline{\sigma}_i \, q(x_i)) + |q(x_i)|^2 \leqslant 2 \, AM + A^2 , \qquad i = 1, \ldots, r .$$

In particular,

$$(3) \qquad 2M \operatorname{Re}(\overline{\sigma}_i \, q(x_i)) \leqslant 2 \, AM + A^2 .$$

Now, from (1), taking $p = q$

$$\overline{\sigma}_j \cdot q(x_j) = - \sum_{i \neq j} (c_i/c_j) \, \overline{\sigma}_i \, q(x_i) ,$$

which gives, using (3)

$$2M \operatorname{Re}(\overline{\sigma}_j \, q(x_j)) = - \sum_{i \neq j} (c_i/c_j) \, 2M \operatorname{Re}(\overline{\sigma}_i \, q(x_i)) \geqslant - B(2 \, AM + A^2)(r - 1)$$

where B denotes the maximum, over all pairs i, j with $i \neq j$, of c_i/c_j. Combining this with (3) gives

$$(4) \qquad 2M \, |\operatorname{Re}(\overline{\sigma}_j \, q(x_j))| \leqslant C_1(2 \, AM + A^2)$$

where C_1 (as $C_2, \ldots$ below) is a constant depending only on f and on P.
From (2), (4) we get

$$(5) \qquad |q(x_i)|^2 \leqslant C_2(2 \, AM + A^2) \qquad i = 1, \ldots, r.$$

Now, since no element of P except zero can vanish at all the points x_i, we have

$$(6) \qquad \|p\| \leqslant C_3 \max_i |p(x_i)| , \qquad p \in P .$$

(Indeed, this is a consequence of the facts that the right side of (6) is a norm for P, and any two norms on a finite dimensional Banach space are equivalent.) Using this, and (5), gives

$$\|q\|^2 \leqslant C_4(2 \, AM + A^2)$$

and this leads, by a simple computation, to

$$A \geq (M^2 + C_4^{-1} \|q\|^2)^{1/2} - M$$

or in terms of f,

$$\|f + q\| \geq (\|f\|^2 + c \|q\|^2)^{1/2}$$

where c is a constant depending only on f and P. This proves (i). As for (ii), observe that in the real-valued case (4) gives

$$2M |q(x_i)| \leq C_1(2 AM + A^2)$$

whence by (6)

$$M \|q\| \leq C_5(2 AM + A^2)$$

which gives

$$A \geq (M^2 + C_5^{-1} M \|q\|)^{1/2} - M$$

$$\|f + q\| \geq (\|f\|^2 + c \|f\| \cdot \|q\|)^{1/2}, \qquad c = C_5^{-1} .$$

Without loss of generality we may suppose $c \leq 1$. Now, the last expression exceeds $\|f\| + (c/3) \|q\|$ if $\|q\| \leq 3\|f\|/c$. If, however, $\|q\| > 3 \|f\|/c$ we have

$$\|f + q\| \geq \|q\| - \|f\| > \|f\| + (1 - (2c/3)) \|q\|$$

so that (ii) holds in any case. ◇

The following corollary, a strong form of Haar's uniqueness theorem, gives an estimate of how fast the distance from a given continuous function f to a function p in a Haar space increases as p recedes from p^*, the best approximation to f.

<u>2.4.6 Corollary</u>. <u>Let</u> Q <u>be a compact Hausdorff space and let</u> P <u>be a Haar subspace of</u> C(Q). <u>If</u> p^* <u>is the best approximation from</u> P <u>to a given</u> $g \in C(Q)$, <u>then for any</u> $p \in P$,

(i) $$\|g - p\| \geq (\|g - p^*\|^2 + a \|p^* - p\|^2)^{1/2} .$$

<u>In the real-valued case we have</u>

(ii) $$\|g - p\| \geq \|g - p^*\| + b \|p^* - p\| ,$$

<u>where</u> a, b <u>are positive constants depending on</u> g <u>and</u> P, <u>but not on</u> p.

Proof. We have only to take $f = g - p^*$, $q = p^* - p$ in the preceding theorem. ◇

2.4.7 Corollary. (De La Vallée Poussin, Freud). Under the same hypotheses on Q, P, and assuming all functions real-valued, the map A which takes each $f \in C(Q)$ into its (unique) best approximation from P satisfies a Lipschitz condition at each point of $C(Q)$, that is, for each $f_o \in C(Q)$ we have

$$\text{(i)} \qquad \|Af - Af_o\| \leq K \|f - f_o\|$$

where K is a constant depending on f_o and P but not on f.

Proof. Write $p_o = Af_o$, $p = Af$. By the preceding corollary, there exists a constant B depending only on f_o and P such that

$$\|p_o - p\| \leq B(\|f_o - p\| - \|f_o - p_o\|) \leq B(\|f_o - f\| + \|f - p\| - \|f_o - p_o\|).$$

Moreover,

$$\|f - p\| \leq \|f - p_o\| \leq \|f - f_o\| + \|f_o - p_o\|$$

and inserting this in the previous inequality gives (i), with $K = 2B$. ◇

2.5 An Explicit Expression for the Distance in Haar Spaces. Let $-1 \leq x_1 < \ldots, x_{n+1} \leq 1$ and let $\{g_1, \ldots, g_n\}$ be a basis for a real Haar n-space P on I. If the determinant

$$D(f; x_1, \ldots, x_{n+1}) = \begin{vmatrix} f(x_1) & f(x_2) & \ldots & f(x_{n+1}) \\ g_1(x_1) & g_1(x_2) & \ldots & g_1(x_{n+1}) \\ \vdots & \vdots & & \vdots \\ g_n(x_1) & g_n(x_2) & \ldots & g_n(x_{n+1}) \end{vmatrix}$$

is expanded in terms of the minors of the first row,

$$D(f; x_1, \ldots, x_{n+1}) = \sum_{i=1}^{n+1} (-1)^{i+1} D_i f(x_i) ,$$

where D_i denotes the minor of the element in position $(1, i)$, we see, from the fact that the determinant vanishes when f is taken to be one of the g_i, that the numbers $(-1)^{i+1} D_i$ can be considered as the masses at the points x_i of a P-annihilating measure. We have seen (Theorem 2.4.4) that the masses of such a measure alternate in sign. If this annihilating measure is denoted by μ, we have therefore

$$\int |d\mu| = \sum_{i=1}^{n+1} |D_i| = \text{abs. val.} \begin{vmatrix} 1 & -1 & \cdots & (-1)^{n+1} \\ g_1(x_1) & g_1(x_2) & \cdots & g_1(x_{n+1}) \\ \vdots & \vdots & & \vdots \\ g_n(x_1) & g_n(x_2) & \cdots & g_n(x_{n+1}) \end{vmatrix} .$$

Inserting in the Duality Theorem (2.3.7) this yields:

2.5.1 Theorem. Let P be a real Haar n-space on I, and let $g_1, \ldots, g_n$ be a basis for P. For any real $f \in C(I)$, $\min_{p \in P} \|f - p\|$ equals the (attained) maximum of

$$J(f; x_1, \ldots, x_{n+1}) = \text{abs. val.} \frac{\begin{vmatrix} f(x_1) & f(x_2) & \cdots & f(x_{n+1}) \\ g_1(x_1) & g_1(x_2) & \cdots & g_1(x_{n+1}) \\ \vdots & \vdots & & \vdots \\ g_n(x_1) & g_n(x_2) & \cdots & g_n(x_{n+1}) \end{vmatrix}}{\begin{vmatrix} 1 & -1 & \cdots & (-1)^{n+1} \\ g_1(x_1) & g_1(x_2) & \cdots & g_1(x_{n+1}) \\ \vdots & \vdots & & \vdots \\ g_n(x_1) & g_n(x_2) & \cdots & g_n(x_{n+1}) \end{vmatrix}}$$

taken over all subsets of $n + 1$ distinct points of I.

Remark. Although this maximum is difficult to compute in practice, observe that each choice of $(x_1, \ldots, x_{n+1})$ gives a lower bound for dist (f, P).

2.5.2 Corollary. For a real-valued $f \in C(I)$, dist $(f, \mathcal{P}_n)$ equals the (attained) maximum over all sets of $n + 2$ distinct points in I of

$$J_a(f;\, x_1, \ldots, x_{n+2}) = \text{abs. val.} \frac{\begin{vmatrix} f(x_1) & f(x_2) & \ldots & f(x_{n+2}) \\ 1 & 1 & \ldots & 1 \\ x_1 & x_2 & \ldots & x_{n+2} \\ \vdots & \vdots & & \vdots \\ x_1^n & x_2^n & \ldots & x_{n+2}^n \end{vmatrix}}{\begin{vmatrix} 1 & -1 & \ldots & (-1)^{n+2} \\ 1 & 1 & \ldots & 1 \\ x_1 & x_2 & \ldots & x_{n+2} \\ \vdots & \vdots & & \vdots \\ x_1^n & x_2^n & \ldots & x_{n+2}^n \end{vmatrix}}$$

<u>Exercise</u>. Compute this maximum when $n = 1$ and a) $f(x) = x^2$, b) $f(x) = x^3$.

<u>Remark</u>. By virtue of the Weierstrass approximation theorem, to be proved later, the above ratio of determinants (for fixed f) must become arbitrarily small for large n, uniformly with respect to the choice of the x_i. It would be of interest to give a direct proof of this fact (which would then give a new proof of the Weierstrass theorem). This seems quite difficult. We have here a piquant illustration of the gulf that exists between the view-points of "best approximation" and "good approximation", and of the limitations of "exact" solutions to problems, when these are too implicit to allow estimations.

<u>2.6 Extremal Signatures</u>. As before, we study the continuous functions C(Q) defined on a compact Hausdorff space Q.

<u>2.6.1 Definitions</u>. A continuous complex-valued function σ, defined on a closed subset F of Q, is called a <u>signature</u> if $|\sigma(x)| = 1$ for $x \in F$. The set F is called the <u>support</u> of σ. A signature σ with support F is termed an <u>extremal signature with respect to the subspace</u> P of C(Q) (briefly, <u>extremal signature</u> or <u>e.s.</u>) if there exists a positive measure ρ whose support is F such that

$$\int p(x)\,\overline{\sigma(x)}\,d\rho(x) = 0 \quad , \qquad \text{all } p \in P ,$$

i.e. $\bar{\sigma}\,d\rho$ annihilates P. Note that no restriction on the dimension of P is imposed and that F may consist of the whole space Q. In practice, however, F is usually finite. Extremal signatures play the role, in the general theory of approximation, which the alternating-sign patterns play in the classical Chebyshev-Haar theory.

2.6.2 Examples.

2.6.2.1 If $Q = [-1, 1]$, $P = \mathcal{P}_n$ and $F = \{x_1, \ldots, x_{n+2}\}$ where $x_1 < x_2 < \ldots < x_{n+2}$, then $\sigma(x_i) = (-1)^i$ defines an extremal signature, and indeed, modulo multiplication by -1, the only e.s. with this support. This follows from Theorem 2.4.4.

2.6.2.2 Let Q = unit square, $P = \text{span}\{1, x, y\}$. Let σ be a signature and define the sets $E^{\pm}(\sigma) = \{x \in Q \mid \sigma(x) = \pm 1\}$. Then σ is an extremal signature (relative to P) if and only if the convex hulls of E^+ and E^- intersect (exercise for the reader). In this example all e.s. can in fact be reduced to three primitive types (as compared with just one in ex. 1). This is discussed in COLLATZ (cf. BROSOWSKI$_1$, SHAPIRO$_4$ for generalizations).

2.6.2.3 $Q = T$ (the unit circle), $P = \text{span}\{1, z, \ldots, z^n\}$, where $z = e^{i\theta}$. Here $\sigma(z) = z^{n+1}$ is an extremal signature for P on T, since

$$\int \bar{z}^{\,n+1} z^{\nu}\, d\theta = 0 \qquad \nu = 0, 1, \ldots, n .$$

The associated positive measure is Haar measure of the circle (i.e. arc length). (Exercise: find other measures that work here.)

2.6.2.4 $Q = T$, $P = \text{span}\{z, \ldots, z^n\}$. If $\sigma(z) = 1$ on $F = \{1, \omega, \ldots, \omega^n\}$, where $\omega = \exp(2\pi i/(n+1))$, σ is an extremal signature, because

$$\sum_{i=0}^{n} (\omega^{\nu})^i = \frac{1 - \omega^{\nu(n+1)}}{1 - \omega^{\nu}} = 0 , \qquad \nu = 1, \ldots, n.$$

The associated positive measure is a measure with equal masses at each point of F. Also, with the same measure, $\sigma(z) = z^{n+1}$, restricted to F, is an e.s. for P.

2.6.2.5 $Q = \overline{D}$ (closed unit disc), $P = A$ the "disc algebra", i.e. the Banach algebra of all complex-valued functions continuous on $\overline{D}$ and holomorphic in the open disc D. If $|a| < 1$, then $\sigma(z) = \beta(z, a) \underset{\text{def.}}{=} (1 - \overline{a}\, z)/(z - a)$ (inverted Blaschke factor) is an extremal signature for P with support T. To see this, we introduce the positive measure $d\rho(z) = |z - a|^{-2}\, d\theta$ $(z = e^{i\theta})$ on T and observe that

$$\int f(z)\, \overline{\sigma(z)}\, d\rho(z) = \int f(z) \left(\frac{z - a}{1 - \overline{a}z}\right) \frac{d\theta}{(z - a)(\overline{z} - \overline{a})} = \int \frac{f(z)}{(1 - \overline{a}z)^2} \frac{d\theta}{\overline{z}} = -i \int \frac{f(z)dz}{(1 - \overline{a}z)^2} = 0$$

by Cauchy's theorem.

<u>Exercise</u>. Let $|a_i| < 1$ and $|b_j| < 1$. Prove that

$$\frac{\beta(z, a_1) \ldots \beta(z, a_m)}{\beta(z, b_1) \ldots \beta(z, b_n)}$$

is an extremal signature for A when $m > n$.

The Main Theorem (Theorem 2.3.2) can be restated as follows in terms of the notion of extremal signature:

<u>Let</u> P <u>be an</u> n-<u>dimensional subspace of</u> C(Q), <u>the continuous functions on the compact Hausdorff space</u> Q. f <u>is orthogonal to</u> P (<u>i.e.</u> $\|f\| \leq \|f + p\|$, $p \in P$) <u>if and only if there exists a set</u> E <u>in</u> Q <u>with at most</u> $n + 1$ (<u>resp</u>. $2n + 1$) <u>points when</u> C(Q) <u>is real (resp. complex), such that</u> $|f|\,|_E = \|f\|$ <u>and</u> $(f/\|f\|)\,|_E$ <u>is an extremal signature for</u> P. Theorem 2.3.4 can be reformulated in a similar fashion, whereby $f/\|f\|$, restricted to the closed subset F of E, is an extremal signature.

The reason for introducing the notion of extremal signature is to emphasize that it is the <u>signa</u> of the annihilating measures, rather than these measures themselves, which play the dominating role in linear problems of best uniform approximation, and thereby also preserve the formal similarity with the classical Chebyshev theory based on the alternating sign pattern.

<u>2.6.3 Applications</u>. (See also RIVLIN & WEISS.)

2.6.3.1 An application of the thus reformulated Theorem 2.3.4 to 2.6.2.3 shows that the (unique) best uniform approximation to z^{n+1} on T by polynomials of degree $\leq n$ is the zero function. (Uniqueness follows from Haar's theorem.) From 2.6.2.4 it is seen that the best approximation to a constant function on T from $P = \text{span}\left\{z, \ldots, z^n\right\}$ also is identically zero (by a change of variable, this is also deducible from the preceding statement) and from the fact that $\sigma(z) = z^{n+1} + 1$ is an extremal signature for P on T we obtain

$$\max_{|z|=1} |z^{n+1} + a_n z^n + \ldots + a_1 z + 1| \geqslant 2 .$$

2.6.3.2 Let us try to find a best approximation on T to $(z - a)^{-1}$, $|a| < 1$, from A, the disc algebra. (Note that we do not know in advance that one exists.) We know from 2.6.2.5 that $\beta(z, a) = (1 - \bar{a} z)/(z - a)$ is an extremal signature for A, so in seeking a best approximation f^*, it is reasonable to expect that f^* satisfies, for a suitable constant λ,

(1) $$\frac{1}{z - a} - f^*(z) = \lambda \frac{1 - \bar{a} z}{z - a} .$$

To make $f^*(z)$ holomorphic at $z = a$, we must choose $\lambda = (1 - |a|^2)^{-1}$ and we get (since $\beta(z, a)$ has modulus one on T)

$$\left\| \frac{1}{z - a} - f^* \right\| = \frac{1}{1 - |a|^2}$$

and

$$f^*(z) = \frac{\bar{a}}{1 - |a|^2} ,$$

i.e. f^* is constant. Since we have found an admissible f^* and λ so that (1) holds, we know that f^* is indeed a best approximation to $\beta(z, a)$.

<u>Exercise</u>. Prove that the best approximation is unique.

2.6.3.3 Let H^1 denote the class of functions g holomorphic in the open unit disc and such that $g_r(\theta) = g(re^{i\theta})$ are bounded in L^1-norm as $r \rightarrow 1$. This is a Banach space with norm $\|g\|_1 = (2\pi)^{-1} \int_0^{2\pi} |g(e^{i\theta})| \, d\theta$. From 2.6.3.2 we easily obtain a bound for $|g(a)|$ when $g \in H^1$ and $|a| < 1$.

If f^* denotes the same function as in 2.6.3.2,

$$|g(a)| = \left| \frac{1}{2\pi i} \int_{|z|=1} \frac{g(z)}{z - a} \, dz \right| = \left| \frac{1}{2\pi i} \int_{|z|=1} g(z) \left[\frac{1}{z - a} - f^*(z) \right] dz \right| \leqslant$$

$$\leq \left| \frac{1}{2\pi i} \int_{|z|=1} |g(z)|\, (1 - |a|^2)^{-1}\, dz \right| = (1 - |a|^2)^{-1}\, \|g\|_1 .$$

This bound is the best possible because the H^1-norm of $(1 - \bar{a}\, z)^{-2}$ is $(1 - |a|^2)^{-1}$. (Note that by $g(z)$, $z \in T$ we mean the radial boundary values, cf. DUREN.)

2.6.4 Theorem. Let g be a complex-valued function continuous on T, the unit circle, and not belonging to A, the class of functions ("disc algebra") continuous in the closed unit disc and holomorphic on the open disc. Then $g^* \in A$ is a best uniform approximation to g (from A) if and only if

(i) $|g(z) - g^*(z)| = \|g - g^*\|$, $z \in T$,

(ii) there exists $h \in H^1$ such that $z(g(z) - g^*(z))\, h(z) \geq 0$ for almost all $z \in T$.

Proof. Sufficiency. If (i) and (ii) hold, then $z(g(z) - g^*(z))\, h(z)\, d\theta$ is a positive measure on T and for any $f \in A$,

$$i \int_T \overline{(g(z) - g^*(z))}\, z(g(z) - g^*(z))\, h(z)\, f(z)\, d\theta = \|g - g^*\|^2 \int_{|z|=1} h(z)\, f(z)\, dz = 0.$$

Thus $(g(z) - g^*(z))/\|g - g^*\|$ is an extremal signature for A and Theorem 2.3.4 can be applied.

Necessity. If $g - g^*$ is orthogonal to A there exists by the same theorem a positive measure ρ with support on some subset F of the set $E(g - g^*)$ where $|g(z) - g^*(z)|$ is maximal, such that

$$\int_F z^n\, \overline{(g(z) - g^*(z))}\, d\rho = 0 \quad \text{for } n = 0, 1, \ldots$$

But now the theorem of F. and M. Riesz (cf. e.g. DUREN, HOFFMAN) tells us that

$$\overline{(g(z) - g^*(z))}\, d\rho = zh(z)\, d\theta \tag{1}$$

for some $h \in H^1$. Thus the support of ρ must be all of T, for otherwise h would vanish on an arc of T, and hence identically, contradicting the assumption $g \notin A$. This implies (i), and after multiplying both sides of (1) by $g(z) - g^*(z)$ we see that also condition (ii) is necessary. ◇

Let Q_1 and Q_2 be compact Hausdorff spaces and let P_1 be an $(m + 1)$-dimensional subspace of (real) $C(Q_1)$ with basis $\varphi_o, \ldots, \varphi_m$ and P_2 an $(n + 1)$-dimensional subspace of (real) $C(Q_2)$ with basis $\psi_o, \ldots, \psi_n$. Here we suppose $\varphi_o \equiv 1$, $\psi_o \equiv 1$. The $(m + 1)(n + 1)$-dimensional subspace of $C(Q_1 \times Q_2)$ with elements

$$p(x, y) = \sum_{i=o}^{m} \sum_{j=o}^{n} a_{ij} \varphi_i(x) \psi_j(y) ,$$

where the a_{ij} are real numbers, shall be denoted by P (P is often called the <u>tensor product</u> of P_1 and P_2).

<u>2.6.5. Theorem</u>. <u>Let</u> $f \in C(Q_1)$, $g \in C(Q_2)$. <u>If</u> f^* <u>is a best (uniform) approximation to</u> f <u>from</u> P_1 <u>and</u> g^* <u>is a best approximation to</u> g <u>from</u> P_2, <u>then</u> $f^*(x) + g^*(y)$ <u>is a best approximation from</u> P <u>to</u> $f(x) + g(y)$ <u>on</u> $Q_1 \times Q_2$.

<u>Proof</u>. It is sufficient to show, denoting by h the function in $C(Q_1 \times Q_2)$ whose value at (x, y) equals $f(x) + g(y) - f^*(x) - g^*(y)$, that sgn h is an extremal signature relative to P on some subset of $E(h) \subset Q_1 \times Q_2$. We know that $\operatorname{sgn}(f - f^*)$ and $\operatorname{sgn}(g - g^*)$ are extremal signatures for P_1 and P_2 respectively, on suitable subsets of $E(f - f^*)$ and $E(g - g^*)$. Thus, defining

$$E_1^{\pm} = E^{\pm}(f - f^*) = \{x \in Q_1 : (f - f^*)(x) = \pm \|f - f^*\|\}$$

$$E_2^{\pm} = E^{\pm}(g - g^*) = \{y \in Q_2 : (g - g^*)(y) = \pm \|g - g^*\|\},$$

there exists a positive measure ρ supported by (some subset of) $E_1^+ \cup E_1^-$ such that

$$\int \varphi_i(x) \operatorname{sgn} (f - f^*)(x) \, d\rho = 0 , \qquad i = 0, \ldots, m,$$

or, equivalently, writing $\rho = \rho^+ + \rho^-$, where ρ^+, ρ^- are positive measures supported by E_1^+, E_1^- respectively,

$$\int \varphi_i(x) \, d\rho^+ = \int \varphi_i(x) \, d\rho^- \qquad i = 0, \ldots, m.$$

Similarly, there exist positive measures μ^+, μ^- with support on E_2^+, E_2^-, respectively, such that

$$\int \psi_j(y)\, d\mu^+ = \int \psi_j(y)\, d\mu^- \qquad j = 0, \ldots, n.$$

Now define the measure σ^+ on $Q_1 \times Q_2$ to be the product measure $\rho^+ \times \mu^+$, similarly $\sigma^- = \rho^- \times \mu^-$, and finally $\sigma = \sigma^+ + \sigma^-$. We have, since h is negative on the support of σ^- and positive on the support of σ^+,

$$\iint \varphi_i(x)\, \psi_j(y) \operatorname{sgn} h(x, y)\, d\sigma^-(x, y) = - \int \varphi_i(x)\, d\rho^-(x) \int \psi_j(y)\, d\mu^-(y) =$$

$$- \int \varphi_i(x)\, d\rho^+(x) \int \psi_j(y)\, d\mu^+(y) = - \iint \varphi_i(x)\, \psi_j(y)\, d\sigma^+(x, y) =$$

$$= - \iint \varphi_i(x)\, \psi_j(y) \operatorname{sgn} h(x, y)\, d\sigma^+(x, y)$$

This shows that $\int \varphi_i(x)\, \psi_j(y) \operatorname{sgn} h(x, y)\, d\sigma = 0$, hence for all $p \in P$, $\int p \operatorname{sgn} h\, d\sigma = 0$ so that sgn h, restricted to a suitable subset of E(h), is an extremal signature relative to P. ◇

2.6.6 Corollary. Let $f(x)$ be continuous for $a \leq x \leq b$, $g(y)$ continuous for $c \leq y \leq d$, and m, n positive integers. Then, among the polynomials of the form $\sum_{i=0}^{m} \sum_{j=0}^{n} a_{ij}\, x^i y^j$, $p^*(x) + q^*(y)$ is a best uniform approximation to $f(x) + g(y)$ on the rectangle $[a \leq x \leq b,\ c \leq y \leq d]$, where p^*, q^* are the best polynomial approximations of degree m, n to f, g on [a, b], [c, d] respectively.

For further literature on this problem, including discussion of uniqueness, see NEWMAN & SHAPIRO$_1$, LAWSON, GROSOF & NEWMAN. We next state a theorem of similar character.

2.6.7 Theorem. Let P denote the class of functions of the form

$$p(x, y) = \sum_{i=0}^{m} \varphi_i(x)\, b_i(y) + \sum_{j=0}^{n} \psi_j(y)\, a_j(x)$$

where the functions $\varphi_i(x)$, $\psi_j(y)$ form bases for P_1, P_2 resp., and $a_j(x)$, $b_i(y)$ denote arbitrary real functions in $C(Q_1)$, $C(Q_2)$ resp. Let $f(x) \in C(Q_1)$ and $g(y) \in C(Q_2)$ be real-valued and let $f^*(x)$ and $g^*(y)$ be best (uniform) approximations to $f(x)$ and $g(y)$ from P_1 and P_2. In the uniform norm on $C(Q_1 \times Q_2)$, a best approximation from P to $f(x)\, g(y)$ is given by

$$p^*(x, y) = f(x)\, g^*(y) + f^*(x)\, g(y) - f^*(x)\, g^*(y)$$

and

$$\|f(x)\, g(y) - p^*(x, y)\| = \|f - f^*\|\, \|g - g^*\| .$$

As the proof is similar to that of Theorem 2.6.5, we leave it as an exercise (cf. SHAPIRO_4), as well as verification of the next assertion.

2.6.8 Corollary.

$$\min_{a_{ij}} \max_{x,y \in I} |x^m y^n - {\sum}' a_{ij} x^i y^j| \geq \max_{x,y \in I} |(2^{-m+1} C_m(x))(2^{-n+1} C_n(y))| = 2^{-m-n+2},$$

where ${\sum}'$ denotes summation over any finite collection of pairs (i, j) with $i \geq 0$, $j \geq 0$, $\min(i - m, j - n) < 0$, and C_ν denotes the Chebyshev polynomial of order ν.

2.6.9 Exercises. a) Let Q', Q'' be a pair of interlocking rectangles in R^2 (i.e. each has precisely one vertex lying inside the other) with corresponding sides parallel. Show that the vertices of Q' are the "plus" points, and the vertices of Q'' the "minus" points, of an extremal signature for the class of polynomials in x, y having degree at most 2 in each variable.

b) Let Q', Q'' be finite subsets of R^n whose convex hulls intersect, and which are minimal with respect to this property. Show that Q' is the set of "plus" points, and Q'' the "minus" points, of an extremal signature for the class of linear polynomials in $x_1, \ldots, x_n$. (For a complete classification of e.s. for linear polynomials, see BROSOWSKI_1).

Remark. For a method of constructing e.s. for polynomials of higher degree in several variables, see SHAPIRO_4. However, apart from a few trivialities, nothing is yet known about this subject, which seems to offer an attractive field for exploration, impinging upon the relatively neglected area of algebraic geometry over the real field.

c) Let $Q = T$ = unit circle, and V a closed subalgebra of C(T) containing ψ, $\psi(e^{it}) = e^{it}$.

(i) Prove that if $V \neq C(T)$, $\bar{\psi}$ is orthogonal to V (here you may use the (Weierstrass) theorem that the polynomials in ψ and $\overline{\psi}$ are dense in C(T)).

(ii) Deduce that if V contains any function which has a non-vanishing Fourier coefficient of negative index, V = C(T) (Wermer maximality theorem; fuller discussion and references in SHAPIRO_6).

2.7 Concluding remarks. In closing this chapter, we may refer the reader wishing to further pursue the preceding topics to SINGER_1 and references there. In this chapter (also in the following) the proofs were based upon convexity, and it should be pointed out that convexity principles have been developed into a powerful methodology for the study of diverse problems in analysis. Many interesting applications will be found in KARLIN & STUDDEN; for detailed treatments of convexity as such, together with applications see ROCKAFELLAR, STOER & WITZGALL.

Chapter 3. The Interpolation Formula and Gaussian Quadrature

3.1 The Interpolation Formula

3.1.1 Theorem (Interpolation Theorem). *Let* P *be an* n*-dimensional subspace of* $C(Q)$*, the linear space of continuous functions defined on the compact Hausdorff space* Q*, and let* L ($\neq 0$) *be a linear functional on* P*. Then, there exist points* $x_1, \ldots, x_r$ *in* Q*, where* $1 \leq r \leq n$ *when* $C(Q)$ *is real and* $1 \leq r \leq 2n - 1$ *when* $C(Q)$ *is complex, and non-zero constants* $a_1, \ldots, a_r$ *such that*

(1) $$Lp = \sum_{i=1}^{r} a_i \, p(x_i) , \qquad p \in P .$$

and

(2) $$\|L\| = \sum_{i=1}^{r} |a_i| .$$

Proof. Because of the finite dimensionality of P there exists an element $p^* \in P$ (called an extremal element for L) such that $\|p^*\| = 1$ and $Lp^* = \|L\|$. Let P_o denote the nullspace of L, i.e. $P_o = \{p \in P: Lp = 0\}$. Now p^* is orthogonal (see the previous chapter) to P_o, because if, for some $p_o \in P_o$, $\|p^* + p_o\| < \|p^*\| = 1$, we should have an element g with $\|g\| < 1$, such that $Lg = \|L\|$, which is impossible. Then, noting that P_o has real dimension $n - 1$ ($2(n - 1)$) when $C(Q)$ is real (complex), we may by the Main Theorem (2.3.2) assert the existence of points $x_1, \ldots, x_r$, $1 \leq r \leq n$ ($1 \leq r \leq 2n - 1$), in $E = \{x \in Q: |p^*(x)| = 1\}$ and positive constants $c_1, \ldots, c_r$, such that

$$\sum_{i=1}^{r} c_i \, \overline{p^*(x_i)} \, p_o(x_i) = 0 \qquad p_o \in P_o .$$

Since $(Lp)p^* - (Lp^*)p \in P_o$ for all $p \in P$ we have

$$Lp \sum_{i=1}^{r} c_i \, |p^*(x_i)|^2 = Lp^* \sum_{i=1}^{r} c_i \, \overline{p^*(x_i)} \, p(x_i) \qquad p \in P .$$

Since $Lp^* = \|L\|$, we obtain (1) by putting $a_i = \|L\| \, c_i \, \overline{p^*(x_i)} \, / \sum_{i=1}^{r} c_i$.

Since $|p^*(x_i)| = 1$, $i = 1, \ldots, r$, it is also obvious that $\sum_{i=1}^{r} |a_i| = \|L\|$. ◇

3.1.2 Remarks. a) Observe that $\operatorname{sgn} a_i = \overline{p^*(x_i)}$ and that if, for some $q^* \in P$, $\|q^*\| = 1$ and $Lq^* = \|L\|$, then, by (1) and (2)

$$Lq^* = \sum_{i=1}^{r} a_i\, q^*(x_i) = \sum_{i=1}^{r} |a_i| ,$$

from which we may conclude that

$$x_i \in E(q^*) = \left\{x \in Q : |q^*(x)| = \|q^*\|\right\}$$

and

$$q^*(x_i) = \overline{\operatorname{sgn} a_i} \qquad i = 1, \ldots, r.$$

Thus, if P is a Haar n-space and $r = n$, the extremal element p^* is unique. An example showing that the bound $r \leqslant n$ no longer holds in the complex case is given in RIVLIN & SHAPIRO. An interesting exception is the space of complex polynomials of degree n in $\left\{|z| \leqslant 1\right\}$, where $n + 1$ points always suffice ($SHAPIRO_2$). A detailed discussion of this case, from a different point of view, may be extracted from GRENANDER & SZEGÖ, Chapter 4.

b) The interpolation formula is also deducible from the Krein-Milman theorem, once the "point functionals" ωL_x, $L_x p = p(x)$, $|\omega| = 1$, defined on P, are identified with the extreme points of the unit ball of the dual space of P. Carathéodory's lemma (2.3.1) then gives a bound for the number of point functionals needed to represent a given functional on P.

c) The interpolation formula may be viewed as a refined form of the Riesz representation theorem for finite-dimensional subspaces of C(Q). It is tempting to try to deduce the Riesz theorem from the interpolation formula, at least in the case where C(Q) is separable. For then, C(Q) contains a nested sequence P_n of finite-dimensional subspaces whose union is dense in C(Q), and if L is a bounded linear functional on C(Q) the interpolation formula tells us that $L|_{P_n}$ is represented by a (discrete) measure μ_n whose total variation is the norm of $L|_{P_n}$, which does not exceed $\|L\|$. By a routine compactness argument, the desired representing measure μ is gotten as a weak* limit of a suitable subsequence μ_{n_i}.

The trouble with this is, the only proof known to us that the unit ball in the space M of bounded measures on Q is compact for the topology on M induced by C(Q) (via the bilinear form $(f, \mu) \rightarrow \int f\, d\mu$) is based upon the knowledge that M is a dual space, i.e. upon the Riesz representation theorem that we are trying to prove. For special spaces Q the desired compactness can be proved directly, e.g. if Q is the unit interval, then (identifying measures with functions of bounded variation) Helly's theorem yields the desired result. It is our feeling that, with some attention to the measure theory, one could prove a fairly general version of Riesz' theorem along the lines just indicated.

d) The interpolation formula may also be seen as a refined version (in a special context) of the Hahn-Banach theorem; namely, it exhibits a norm-preserving extension to C(Q) of a bounded linear functional defined on a finite dimensional subspace thereof (general discussion in VORONOVSKAYA).

3.2 Applications

3.2.1 Polynomials in the unit disc. Let P be the space of complex polynomials of degree n, normed by $\|p\| = \max_{z \in T} |p(z)|$, where T is the unit circle $= \{z: |z| = 1\}$, and α a complex number, $|\alpha| \geq 1$, and consider the linear functional

$$Lp = p'(\alpha) , \qquad p \in P .$$

Then, by the Interpolation Theorem,

$$p'(\alpha) = \sum_{i=1}^{r} a_i \, p(z_i) \tag{1}$$

where $z_i \in T$, $r \leq 2n + 1$, and $\sum |a_i| = \|L\|$. If $r \leq n$, there exists a non-null $p_1 \in P$ that is zero at n distinct points of T among which are $z_1, \ldots, z_r$. If α coincided with one of those n points, then, by (1), $p_1(z)$ should have a double zero at $z = \alpha$, which is a contradiction (to $p_1(z) \not\equiv 0$). Also, when α does not coincide with any of the n distinct zeroes of $p_1(z)$ we get a contradiction, because by a theorem of Gauss (POLYA & SZEGÖ, p. 59) the zeroes of p_1' lie in the convex hull of the zeroes of p_1, whereas (1) implies that $p_1'(\alpha) = 0$. Thus $r \geq n + 1$ and the extremal polynomial, $p^*(z)$ ($\|p^*\| = 1$, $Lp^* = \|L\|$), must attain its maximum modulus at not less than $n + 1$ points of T, i.e., the equation $t(\theta) = 1 - |p^*(e^{i\theta})|^2 = 0$ has at least $n + 1$ distinct roots. But t is a non-negative trigonometric polynomial, hence each root is at least double. It is a simple exercise to show this implies t vanishes identically, i.e., $|p^*(z)| = 1$ when $z \in T$. Hence (exercise!) $p^*(z)$ must be of the form ωz^ν for some ν, $0 \leq \nu \leq n$, where $|\omega| = 1$. Clearly the choice $\nu = n$ gives the largest (in absolute value) derivative at $z = \alpha$, so we have proved

<u>3.2.2 Theorem</u>. <u>Let</u> p <u>be any complex polynomial of degree</u> n, $\|p\| = \max_{|z|=1} |p(z)|$, <u>and</u> $|\alpha| \geq 1$. <u>Then</u> $|p'(\alpha)| \leq n|\alpha|^{n-1} \|p\|$.

<u>Exercises</u>. a) Prove that a complex polynomial of degree n that (restricted to T) attains its maximum modulus at $n + 1$ points of T is of the form cz^ν, $0 \leq \nu \leq n$.

b) Let p be a complex polynomial of degree n. Prove that for $|\alpha| \geq 1$, $|p(\alpha)| \leq |\alpha|^n \max_{|z|=1} |p(z)|$.

<u>3.2.3 Polynomials on I</u>. Let $P = \mathcal{P}_n$, the real algebraic polynomials of degree n, normed by $\|p\| = \max_{x \in I} |p(x)|$ (as usual, $I = [-1, 1]$), and for a given x_0, $x_0 \geq 1$, define

$$Lp = p'(x_0) , \qquad p \in \mathcal{P}_n$$

Then, by 3.1.1 (1)

$$p'(x_0) = \sum_{i=1}^{r} a_i \, p(x_i) \tag{1}$$

where $-1 \leq x_1 < \ldots < x_r \leq 1$ and $r \leq n + 1$. If $r \leq n$ there exists a non-null $p_1 \in \mathcal{P}_n$ vanishing at n distinct points of I, among which are $x_1, \ldots, x_r$. For obvious reasons this contradicts (1). Thus $r = n + 1$ and the extremal polynomial p^* attains its largest modulus at $m \geq n + 1$ points. If this occurs for $m \geq n + 2$ points, then $p^*(x) =$ const., which is clearly impossible. Thus $m = n + 1$, and by the proof of Corollary 2.2.3, p^* is a constant multiple of the Chebyshev polynomial C_n. Hence

(2) $$|p'(x_o)| \leq |C_n'(x_o)| \max_{|x|\leq 1} |p(x)|$$

for all $p \in \mathcal{P}_n$ and $x_o \geq 1$. <u>In particular</u>, $|p'(1)| \leq n^2\|p\|$.

<u>Remark</u>. Since $p'(x_o) = \sum_{i=1}^{n+1} a_i\, p(x_i)$, where $\sum_{i=1}^{n+1} |a_i| = \|L\|$, we have in fact the stronger result

$$|p'(x_o)| \leq |C_n'(x_o)| \max_{1\leq i\leq n+1} |p(x_i)| , \qquad \text{all } p \in \mathcal{P}_n$$

where $x_i = \cos (i - 1)\pi/n$.

<u>Exercise</u>. For $p \in \mathcal{P}_n$, prove that

$$|p(x_o)| \leq |C_n(x_o)| \max_{1\leq i\leq n+1} |p(x_i)|$$

if $|x_o| \geq 1$, where $x_i = \cos (i - 1)\pi/n$.

To handle trigonometric polynomials we need a corresponding lemma. Certain very basic facts about trig. polys. are here taken for granted, see e.g. NATANSON.

<u>3.2.4 Lemma</u>. <u>Let</u> t <u>be a trigonometric polynomial of degree</u> n <u>and suppose</u> $|t(\theta_i)| = \|t\|$ <u>for</u> $0 \leq \theta_1 < \ldots < \theta_r < 2\pi$. <u>If</u>

a) $r = 2n$, <u>then</u> $t(\theta) = \|t\| \cos (n\theta + \alpha)$

b) $r = 2n + 1$, <u>then</u> $t(\theta) \equiv \|t\|$.

The method of proof for a) is the same as in the polynomial case (see the proof of 2.2.3) so the details are left to the reader.

In case b), $t'(\theta) = 0$ at $2n + 1$ points, and a trigonometric polynomial of degree n with $2n + 1$ zeroes vanishes identically.

<u>3.2.5 Derivative of a trigonometric polynomial</u>. Take P to be $\mathcal{T}_n$, the $2n + 1$-dimensional space of trigonometric polynomials of degree at most n, and for a fixed θ_o, $0 \leq \theta_o < 2\pi$, define

$$Lt = t'(\theta_o) , \qquad t \in \mathcal{T}_n.$$

Then, by 3.1.1 (1),

(1) $$t'(\theta_o) = \sum_{i=1}^{r} a_i \, t(\theta_i) ,$$

where $0 \leq \theta_1 < \dots < \theta_r < 2\pi$ and $r \leq 2n + 1$. Now, if $r < 2n$, we can choose a non-null polynomial $t_1 \in \mathcal{T}_n$ with $2n$ distinct zeroes, among which are $\theta_1, \dots, \theta_r$ and $\theta = \theta_o$. But this leads to a contradiction, since then, by (1), $t_1(\theta)$ has a double zero at $\theta = \theta_o$, and an element of $\mathcal{T}_n$ having $2n + 1$ zeroes must vanish identically. Also, $r = 2n + 1$ is impossible, since then the extremal polynomial $t^*(\theta)$ should attain its maximum modulus at least at $2n + 1$ points, and hence by Lemma 3.2.4 be constant, and as in the algebraic case this leads to a contradiction. Thus $r = 2n$, and by 3.2.4, $t^*(\theta) = \cos (n\theta + \alpha)$ for some α. Hence,

$$|(t^*)'(\theta_o)| = n \, |\sin (n\theta_o + \alpha)| \leq n .$$

Since $(t^*)'(\theta_o)$ shall be positive and as large as possible, we must have $n\theta_o + \alpha \equiv$ $\equiv - (\pi/2)(\text{mod } 2\pi)$, which gives $t^*(\theta) = \sin n(\theta - \theta_o)$. We have thus proved:

<u>3.2.6 Theorem (Bernstein inequality)</u>. <u>For</u> $t \in \mathcal{T}_n$, α <u>real</u>,

$$|t'(\alpha)| \leq n \max_{\theta} |t(\theta)|$$

<u>equality holding if and only if</u> $t(\theta) = c \sin n(\theta - \alpha)$ <u>for some constant</u> c.

<u>3.2.7 Remark</u>. The "extremal points" for $t^*(\theta)$ in 3.2.5 are $\theta_i = \theta_o + ((2i-1)\pi/2n)$, $i = 1, \dots, 2n$. This implies the interpolation formula, valid for all $t \in \mathcal{T}_n$

(1) $$t'(\theta) = \sum_{i=1}^{2n} a_i \, t(\theta + ((2i - 1)\pi/2n))$$

where the a_i are real numbers satisfying $\sum_{i=1}^{2n} |a_i| = n$. This is, apart from the explicit determination of the a_i, an identity discovered by M. Riesz. From (1) we deduce

$$|t'(\theta)| \leq n \max_{1 \leq i \leq 2n} |t(\theta + ((2i - 1)\pi/2n))| .$$

Moreover, since $|t'(\theta)| \leq \sum_{i=1}^{2n} |a_i| \, |t(\theta + ((2i - 1)\pi/2n))|$ we obtain, integrating,

$$\int_0^{2\pi} |t'(\theta)|\, d\theta \leq n \int_0^{2\pi} |t(\theta)|\, d\theta ,$$

valid for all $t \in \mathcal{T}_n$ (Bernstein inequality in L^1). It is easy also to deduce the analogous L^p inequality, as well as pick out the cases of equality.

The procedure just employed can be carried out for any linear functional on $\mathcal{T}_n$, and leads to the result: If A is a linear operator from $\mathcal{T}_n$ to $\mathcal{T}_n$ that commutes with translations, its norm as an operator from L^1 to L^1 does not exceed its norm as an operator from C to C.

3.2.8 Exercises. a) Prove the last assertion. (Hint: show that A can be represented in the form $(Af)(u) = LS_u f$ where S_u denotes translation, $(S_u f)(\theta) = f(\theta + u)$, and L is a linear functional on T_n. Apply the interpolation formula to L.).

b) Compute the coefficients a_i in M. Riesz' identity.

c) Prove that a bounded linear transformation from $\mathcal{T}_n$ into itself that commutes with translations can be extended with preservation of norm to a bounded linear transformation from C(T) into itself that commutes with translations. Can you generalize this to any other Banach spaces, and to subspaces that needn't be finite dimensional?

d) Let A be a bounded linear operator from C(T) into itself that commutes with translations, and of norm M. Prove that $\|Af\|_p \leq M \|f\|_p$ for all $f \in C(T)$, if $1 \leq p < \infty$ (this is related to "Fourier multipliers", cf. HÖRMANDER$_2$).

e) Consider the linear functional L: $Lp = p(2)$ for $p \in \mathcal{P}_2$, where $\mathcal{P}_2$ is considered as a subspace of C(I). Exhibit a Hahn-Banach (i.e. norm-preserving) extension of L from $\mathcal{P}_2$ to all of C(I). Is this extension unique? Solve the analogous problem for $\mathcal{P}_n$, and also for the functionals $Lp = p'(1)$, $Lp = \int_{-1}^{1} p(x)dx$, and $Lp = \int_{-1}^{1} p(x) \sin 2\pi x\, dx$.

f) Consider, on the space $\mathcal{P}_n$ (now considered as polys. with complex coefficients, defined in $\{|z| \leq 1\}$), the linear functional $Lp = p(0) + \lambda p'(0)$ (λ is some fixed complex number). Prove there is a unique $p^* = p_n^*$ of norm 1 such that

$Lp^* = \|L\|$. Prove also the following: if $0 \leq \theta_1 < \theta_2 \leq 2\pi$, the number of points on the arc joining $e^{i\theta_1}$ to $e^{i\theta_2}$ where $|p_n^*(e^{i\theta})| = 1$ equals $((\theta_2 - \theta_1)/2\pi)n + o(n)$.

3.2.9 Derivatives of algebraic polynomials. We can make use of Bernstein's inequality to derive bounds for derivatives of real algebraic polynomials.

If $p \in \mathcal{P}_n$, then by the substitution $x = \cos\theta$, $p(x)$ is transformed into a trigonometric polynomial $t(\theta)$ of degree n:

$$p(x) = p(\cos\theta) = t(\theta) ,$$

$$p'(x) = -t'(\theta)/\sin\theta .$$

Since $\|p\| = \max_{|x|\leq 1} |p(x)| = \max_{0\leq\theta\leq 2\pi} |t(\theta)|$, Bernstein's inequality gives

$$(1) \qquad |p'(x)| \leq \frac{n\|p\|}{\sqrt{1-x^2}} , \qquad |x| \leq 1 .$$

which is valid for all $p \in \mathcal{P}_n$.

The inequality (1) is clearly not sharp on all of I. By a method due to Schur it is easy to derive a uniform bound for the derivative on the closed interval.

3.2.9.1 Theorem. (Markov's inequality). For $p \in \mathcal{P}_n$,

$$(1) \qquad \max_{|x|\leq 1} |p'(x)| \leq n^2 \max_{|x|\leq 1} |p(x)| .$$

Proof. Let $\{x_i\}$ denote the zeroes of the Chebyshev polynomial C_n, $x_i = \cos(2i-1)\pi/2n$, $i = 1, \ldots, n$. For $|x| \leq x_1$, we have $(1-x^2)^{1/2} \geq (1-x_1^2)^{1/2} = \sin\pi/2n \geq 1/n$, and applying 3.2.9 (1), the theorem is proved for such x. To verify the inequality for the remaining points we apply Lagrange's interpolation formula to $p'(x)$ at the points x_i and obtain

$$p'(x) = \sum_{i=1}^{n} p'(x_i) \frac{C_n(x)}{(x-x_i)\, C_n'(x_i)} = \sum_{i=1}^{n} p'(x_i) \frac{(-1)^{i-1}(1-x_i^2)^{1/2}\, C_n(x)}{n(x-x_i)} .$$

Hence

$$|p'(x)| \leq \frac{1}{n} \sum_{i=1}^{n} \left| \frac{C_n(x)}{x - x_i} \right| \max_{1 \leq i \leq n} |p'(x_i)(1 - x_i^2)^{1/2}|$$

$$\leq \|p\| \sum_{i=1}^{n} \left| \frac{C_n(x)}{x - x_i} \right| , \qquad \text{by 3.2.9 (1),}$$

$$= \|p\| \cdot \left| \sum_{i=1}^{n} \frac{C_n(x)}{x - x_i} \right|$$

(since, for $|x| \geq x_1$, the numbers $(x - x_i)$ all have the same sign)

$$= \|p\| \, |C_n'(x)| .$$

and since $|C_n'(\cos \theta)| = n \, |\sin n\theta / \sin \theta| \leq n^2$, the proof is complete. ◇

Collecting our inequalities we arrive at the following result.

<u>3.2.10 Theorem</u>. <u>For</u> $p \in \mathcal{P}_n$, <u>with real coefficients, and letting</u> $\|p\| = \max_{|x| \leq 1} |p(x)|$,

(i) $$|p'(x)| \leq n \min \left\{(1 - x^2)^{-1/2}, n\right\} \|p\| , \qquad |x| \leq 1$$

(ii) $$|p'(x)| \leq |C_n'(x)| \cdot \|p\| , \qquad |x| \geq 1 ,$$

<u>where</u> C_n <u>denotes the Chebyshev polynomial of degree</u> n.

<u>Remarks</u>. Another way to prove the inequality $|p'(x)| \leq n^2 \|p\|$ for $p \in \mathcal{P}_n$, $|x| \leq 1$, is to consider (for fixed x) the extremal problem of maximizing $p'(x)$ subject to the constraint $\|p\| = 1$, with the aid of the interpolation formula. It is easy to see that the functional $f'(x)$ cannot be interpolated with less than n nodal points; hence (by our usual argument) it is enough to prove the desired inequality for $p \in \mathcal{P}_n$ which attain their maximum modulus on $[-1, 1]$ at $n + 1$ points (i.e. the Chebyshev polynomial of degree n), or at n points. By a procedure similar to that we used in Chapter 2 (but rather more tedious) it is possible to explicitly find all polynomials of the latter type, and so verify the desired inequality (cf. V. MARKOV).

It seems that morally one should be able to infer the desired inequality from the case $x = 1$, but we know of no trivial way to do this.

3.3 Representation of strictly positive functionals, and Gaussian quadrature

A linear functional L on real-valued C(Q) is called positive if $p(x) \geq 0$, all $x \in Q$, implies $Lp \geq 0$. For a positive functional L defined on an n-dimensional subspace P of C(Q) containing $\underline{1}$ (we use the symbol $\underline{1}$ to denote the function identically equal to 1) the coefficients a_i in the interpolation formula are all positive. For, $\underline{1}$ is then an extremal function, hence

$$L\underline{1} = \sum a_i = \|L\| = \sum |a_i| .$$

This situation will now be refined for the case $P = \mathcal{P}_n$. For simplicity, we shall work on the interval $I = [-1, 1]$.

First, we define a strictly positive linear functional L on $\mathcal{P}_n$ as a linear functional with the property that $p \in \mathcal{P}_n$, $p(x) \geq 0$ for all x, and $p(x) \not\equiv 0$ implies $Lp > 0$. We shall see that for a strictly positive functional on $\mathcal{P}_n$, the number r of evaluation points needed in the interpolation formula can be chosen considerably less than n. Let us first consider some obvious limitations for this problem, i.e. lower bounds for r (here understood to be the smallest r for which an interpolation formula for L is possible). If n is even, then $r \geq (n/2) + 1$, because for $r \leq n/2$, $p_1(x) = \prod_{i=1}^{r} (x - x_i)^2 \in \mathcal{P}_n$, so if an interpolation formula with nodal points x_i existed, we should have $Lp_1 = \sum_{i=1}^{r} a_i \, p_1(x_i) = 0$, which contradicts the strict positivity of L. For the same reason, $r \geq (n + 1)/2$ for n odd. In fact, it shall be proved that $r = (n + 1)/2$ nodal points is sufficient for odd n.

Now, let L be a strictly positive functional on $\mathcal{P}_n$, and define $m = [n/2] + 1$, i.e., the smallest integer $> n/2$. We can now (by the standard Gram-Schmidt method) form the orthogonal polynomials with respect to L, i.e., polynomials

$$p_o(x) = 1, \; p_1(x) = x + a_{1,o}, \; \ldots, \; p_m(x) = x^m + a_{m,m-1} \, x^{m-1} + \ldots + a_{m,o},$$

satisfying $L(p_i p_j) = 0$, $0 \leq i < j \leq m$. (When n is even, $m + (m - 1) = n + 1$, so when n is even we will assume tacitly that L extends as a positive linear functional to $\mathcal{P}_{n+1}$.) The orthogonal polynomials have a simple property that will be of frequent use in the sequel.

3.3.1 Lemma. The L-orthogonal polynomial $p_\nu(x)$ of degree ν, $\nu \leq m$, has exactly ν distinct zeroes in the interior of I.

Proof. If $p_\nu(x)$ has $\leq \nu - 1$ zeroes, then there exists a linear combination q of the polynomials $p_o(x), \ldots, p_{\nu-1}(x)$, such that $p_\nu(x)\, q(x) \geq 0$ on I. Thus, $L(p_\nu q) > 0$, which contradicts the orthogonality of the polynomials. (To prove that the zeroes lie in the interior of the interval is left as an exercise.) ◇

Exercise. Prove that the zeroes of the orthogonal polynomials $p_{\nu-1}$ and p_ν $(\nu \geq 2)$ separate each other.

3.3.2 Lemma. Let $L, L_1, \ldots, L_m$ be $m + 1$ continuous linear functionals defined on some linear space X. If $L_1(x) = \ldots = L_m(x) = 0$ implies $L(x) = 0$, then L is a linear combination of the functionals $L_1, \ldots, L_m$.

This is elementary linear algebra, and we omit the proof.

3.3.3 Theorem. Let L be a strictly positive functional defined on $\mathcal{P}_n \subset C(I)$. Let $m = [n/2] + 1$ and denote by p_ν, $\nu = 0, \ldots, m$, the orthogonal polynomials with respect to L. Let x_i, $-1 < x_1 < \ldots < x_m < 1$, denote the zeroes of p_m. Then

(i) For suitably chosen positive numbers a_i

$$(1) \qquad Lp = \sum_{i=1}^{m} a_i\, p(x_i)\ , \qquad \text{all } p \in \mathcal{P}_n .$$

(ii) Moreover, if n is odd, and

$$Lp = \sum_{i=1}^{r} b_i\, p(y_i)\ , \qquad \text{all } p \in \mathcal{P}_n ,$$

where $\{b_i\}$ are real numbers and $-1 < y_1 < \ldots < y_r < 1$, then $r \geq m$, and if $r = m$ then $y_i = x_i$ and $b_i = a_i$, $i = 1, \ldots, m$.

Remark. In special cases L can of course be represented using less than m point evaluations.

Proof. (i) Existence. Let $p \in \mathcal{P}_n$, $p(x_i) = 0$, $i = 1, \ldots, m$. Then $p(x) = q(x)\, p_m(x)$, where the degree of q is less than m. Thus, by the orthogonality, $Lp = 0$. An application of Lemma 3.3.2 now gives the required formula, for the moment with real a_i. To prove the positivity of the a_i, take

(2) $$p(x) = (p_m(x)/(x - x_i))^2 .$$

Then $p \in \mathcal{P}_n$ and $Lp > 0$. But

(3) $$Lp = a_i \, (p_m'(x_i))^2$$

and since p_m has only simple zeroes, $a_i > 0$.

(ii) Uniqueness. Assume n is odd, and

(4) $$Lp = \sum_{i=1}^{r} b_i \, p(y_i) , \qquad \text{all } p \in \mathcal{P}_n ,$$

where we suppose first $r \leq m - 1$. Then $2r \leq 2m - 2 = n - 1$, so that $p(x) = \prod_{i=1}^{r}(x-y_i)^2 \in \mathcal{P}_n$. Inserting this polynomial in (4) gives a contradiction to the strict positivity of L. Thus $r \geq m$. Now, if $r = m$, all the b_i must be non-zero. Apply (4) to the polynomial $q(x) = p_m(x) \prod_{i=2}^{m} (x - y_i)$ (observe that $q \in \mathcal{P}_n$). $Lq = 0$ because of the orthogonality, and the right side of (4) becomes $p_m(y_1) \prod_{i=2}^{n} (y_1 - y_i)$, which shows that y_1 is a zero of p_m. In the same way it can be shown that all the y_i are zeroes of $p_m(x)$, and in view of how the points were ordered, $y_i = x_i$, $i = 1, \ldots, m$. Finally, applying (4) and (1) to $p_m(x)(x - x_i)^{-1}$, we find that $b_i = a_i$. ◇

Remark. (2) and (3) yield a formula for a_i; another such is (in a slightly abusive notation)

$$a_i = p_m'(x_i)^{-1} \, L(p_m(x)/(x - x_i))$$

If the polynomials $p_\nu(x)$, $\nu = 0, \ldots, m$ are normalized so that $L(p_\nu^2) = 1$, rather than being monic, and $p_\nu(x)$ has leading coefficient c_ν, then it is not hard to show

(5) $$a_i = \frac{c_m}{c_{m-1}} \, \frac{1}{p_m'(x_i) \, p_{m-1}(x_i)} , \qquad i = 1, \ldots, m .$$

An immediate corollary of Theorem 3.3.3 is a (generalized) Gaussian quadrature formula: Let σ be a positive measure on I. Then there exist points x_i,

$-1 < x_1 < \ldots < x_m < 1$, <u>and positive constants</u> a_i, <u>such that</u>

$$\int p(x)\, d\sigma = \sum_{i=1}^{m} a_i\, p(x_i)$$

<u>for all algebraic polynomials of degree not exceeding</u> $2m - 1$.

<u>Examples</u>. a) Choosing $d\sigma = dx$, the orthogonal polynomials $p_\nu(x)$, are the <u>Legendre polynomials</u> on I, so if $x_1, \ldots, x_m$ denote the zeroes of the mth Legendre polynomial $p_m(x)$ (normalized so that $\int_{-1}^{1} p_m(x)^2 dx = 1$), we have the classical Gaussian quadrature formula (by means of (5)):

$$(6) \qquad \int_{-1}^{1} f(x)\, dx \cong \sum_{i=1}^{m} \left(\frac{2}{m p_m'(x_i)\, p_{m-1}(x_i)} \right) f(x_i) \ ,$$

which is exact when f is a polynomial of degree $2m - 1$. If we want a quadrature formula for an integral $\int_a^b f(x)dx$ to be exact for polynomials of as high degree as possible, the evaluation points should be the zeroes of the Legendre polynomial on $[a, b]$ of the appropriate degree. For an error estimate of formula (6), see KRYLOV, p. 109.

b) Take $n = 2m-1$ and let $Lp = \int_{-1}^{1} p(x)(1 - x^2)^{-1/2}dx$. Here the orthogonal polynomials p_ν are the Chebyshev polynomials, and hence $x_i = \cos((2i - 1)\pi/2m)$, $i = 1, \ldots, m$. A computation of the coefficients a_i by means of (5) gives $a_i = \pi/m$, $i = 1, \ldots, m$, i.e., each evaluation is <u>equally weighted</u> in this case.

<u>Exercises</u>. a) Prove that the error in the formula (6) tends to zero as $m \to \infty$ for every $f \in C(I)$. (Deduce this from the following general principle: if L_n are uniformly bounded linear functionals on $C(I)$, and $L_n f \to Lf$ for a dense set of f, where L is some bounded linear functional, then $L_n f \to Lf$ for all $f \in C(I)$.)

b) Prove, for $f \in C[0,1]$

$$(*) \qquad \lim_{n \to \infty} \left[\sum_{k=0}^{n} (-1)^k \binom{n}{k} f(k/n) \right] 2^{-n} = 0 \ .$$

Show that (*) holds also for any f of bounded variation.

As a final application of Theorem 3.3.3 we consider the following problem.

Given a positive measure σ on $[0, 1]$ find $p \in \mathcal{P}_n$, non-negative on $[0, 1]$, such that $\int xp(x)d\sigma$ attains a minimum (maximum) subject to the constraint $\int p(x)d\sigma = 1$. (This extremal problem is important for certain applications, cf. NEWMAN & SHAPIRO$_2$.) We give the solution in the case n even. Define $m = (n/2) + 1$ and construct the polynomials p_ν, $\nu = 0, 1, \ldots, m$, orthogonal with respect to σ. Denote the m zeroes of $p_m(x)$ by x_i, $0 < x_1 < \ldots < x_m < 1$. Then, since $xp \in \mathcal{P}_{n+1}$, the quadrature formula gives, for suitable positive numbers a_i,

$$\int xp(x)\, d\sigma = \sum_{i=1}^{m} a_i\, x_i\, p(x_i) \geq x_1 \sum_{i=1}^{m} a_i\, p(x_i) = x_1 \int p(x)\, d\sigma = x_1 .$$

Equality may be attained here, as we see from the choice of $p(x) = (p_m(x)/(x - x_1))^2$, which has degree n. Since a similar argument works for the maximum of $\int xp(x)\, d\sigma$, we can state:

3.3.4 Theorem. The minimum (maximum) of $\int xp(x)\, d\sigma$ over all non-negative polynomials p of degree not exceeding 2k, subject to the normalization $\int p(x)\, d\sigma = 1$, is the smallest (largest) root of p_{k+1}, the orthogonal polynomial of degree $k + 1$ belonging to σ.

Remark. In a similar manner one can get estimates for the higher moments $\int x^r p(x)d\sigma$, but the exact determination of the extrema is now more difficult. For detailed discussion of this problem and generalizations thereof, see KREIN, KARLIN & STUDDEN.

In concluding this chapter, we remark (what is doubtless evident) that we have only scratched the surface of a large literature. Concerning orthogonal polynomials and quadrature formulas, see SZEGÖ, FREUD$_1$. Concerning extremal problems for polynomials, see RAHMAN where also further references to relevant work of S. Bernstein, Fekete, M. Riesz, F. Riesz, Rogosinski, Voronovskaya, and others may be found.

Chapter 4. Best Approximation and Extremal Problems in Other Norms

4.1 Strict and Uniform Convexity. Let X be a normed linear space and Y a closed subspace of X. Given x_o, not in Y, does there exist an element of best approximation to x_o from Y, i.e., is

$$\operatorname{dist}(x_o, Y) \underset{\text{def.}}{=} \inf_{y \in Y} \|x_o - y\|$$

an attained minimum?

In general, the answer is no, but one can give several conditions that assure an affirmative answer. Theorem 2.1.1 gave one such condition, viz. finite-dimensionality of Y. That some restrictions must be placed on X in the infinite-dimensional case can be seen from the following simple example.

Take $X = c_o$, i.e. the elements of X are infinite sequences $x = (x_1, x_2, \ldots)$ where $x_i \to 0$, and $\|x\| = \max |x_i|$. Let Y be the closed subspace of X defined by $Y = \{y \in X: \sum_{i=1}^{\infty} 2^{-i} y_i = 0\}$. In Y there is never a best approximation to any element outside Y, for if $x \in X$ satisfies $\sum_{i=1}^{\infty} 2^{-i} x_i = \alpha > 0$, then, for any $y \in Y$,

$$\|x - y\| > \sum_{i=1}^{\infty} 2^{-i}(x_i - y_i) = \sum_{i=1}^{\infty} 2^{-i} x_i = \alpha ,$$

where the strict inequality holds because $x_i - y_i \to 0$. On the other hand, it is easily seen that for any $\varepsilon > 0$ there exists an element $y_\varepsilon \in Y$ such that $\|x - y_\varepsilon\| < \alpha + \varepsilon$, i.e. $\operatorname{dist}(x, Y) = \alpha$.

Moreover, assuming the existence of a best approximation, uniqueness of this element can only be expected when further conditions are imposed on X, as we have seen in connection with Haar's theorem. The additional conditions assuring existence and uniqueness of the best approximation from a closed subspace that we are going to consider are "convexity" conditions.

A normed space X is said to be strictly convex, or strictly normed, if $\|x + y\| = \|x\| + \|y\|$ implies the existence of some $z \in X$ such that $x = \alpha z$, $y = \beta z$, where $\alpha, \beta \geqslant 0$. This means that (the surface of) the unit sphere U (or any sphere) contains no line segment. The strict convexity is also equivalent to the condition that every supporting hyperplane of U intersects U at only one point. (However, there may exist several supporting hyperplanes through a point of U.) Examples of strictly convex normed spaces are the L^p spaces for $1 < p < \infty$. This is inferred from the conditions for equality to hold in the Minkowski inequality. From simple examples it is also seen that e.g. the spaces c_o, ℓ^1, ℓ^∞, L^1 and L^∞ are not strictly convex. For further discussion of strict convexity see e.g. DAY_1. The significance for our purposes of this concept is that strict convexity assures uniqueness of the element of best approximation if such an element exists.

4.1.1 Theorem. If X is a strictly convex normed space and Y a closed subspace then every element in X has at most one element of best approximation in Y.

Proof. Let y_1 and y_2 in Y be best approximations to an element x_o outside Y. Writing δ for dist (x_o, Y) we get

$$\delta \leq \|x_o - \frac{y_1 + y_2}{2}\| = \| \frac{x_o - y_1}{2} + \frac{x_o - y_2}{2}\| \leq \frac{1}{2} \|x_o - y_1\| + \frac{1}{2} \|x_o - y_2\| = \delta ,$$

and so, for some z, $x_o - y_1 = \alpha z$ and $x_o - y_2 = \beta z$. But since $\|x_o - y_1\| = \|x_o - y_2\|$, $\alpha = \beta$. ◇

A somewhat stronger convexity condition that also assures existence of best approximations in Banach spaces is that of uniform convexity. A normed space is said to be uniformly convex if to every ϵ with $0 < \epsilon \leq 2$ there exists a $\delta(\epsilon) > 0$ such that $\|x\| = \|y\| = 1$ and $\|(x + y)/2\| > 1 - \delta(\epsilon)$ implies $\|x - y\| < \epsilon$. That is, when the midpoint of the line segment joining two points on the unit sphere (or, of course, on any sphere) approaches the sphere then also the endpoints of the segment must approach each other.

Clearly, a uniformly convex space is strictly convex. The converse is not true, but we have at least the following:

Exercise. Every finite-dimensional strictly convex normed space is uniformly convex.

A fundamental property of uniformly convex Banach spaces (due to Milman) is that such spaces are reflexive.

The function

$$\delta(\epsilon) = \inf (1 - \|(x + y)/2\|) : \|x\| = \|y\| = 1, \|x - y\| = \epsilon$$

is called the modulus of uniform convexity of the space. For Hilbert spaces it is easily seen from the parallelogram law that the modulus of uniform convexity equals

$$\delta_h(\epsilon) = 1 - (1 - (\epsilon^2/4))^{1/2} .$$

This modulus turns out to be the largest possible in spaces with real scalars, i.e. in such a space we always have

$$(1) \qquad \delta(\epsilon) \leqslant \delta_h(\epsilon)$$

(NORDLANDER). Conversely, a theorem by Day (DAY_2) states that if equality in (1) holds for all ϵ, $0 < \epsilon \leqslant 2$, then the space is an inner product space. Thus the relation $\delta(\epsilon) = \delta_h(\epsilon)$ characterizes such spaces. Further examples of uniformly convex Banach spaces are the spaces ℓ^p and L^p for $1 < p < \infty$ (see HANNER, KÖTHE) so the following general theorem applies to those spaces.

4.1.2 Theorem. Let X be a uniformly convex Banach space, K a closed convex subset of X and $\delta = \inf_{x \in K} \|x\|$. Then, any sequence $\{x_n\}$ of elements in K satisfying $\|x_n\| \to \delta$ converges, and in particular, K contains a unique element of minimal norm.

4.1.3 Corollary. Under the same assumptions as in the theorem (in particular, if K is a closed subspace) there exists to any element x_o a corresponding unique element y_o in K satisfying $\|x_o - y_o\| = \inf_{y \in K} \|x_o - y\|$.

Proof of Theorem 4.1.2. If $\delta = 0$ the theorem is immediate since K is closed and X is complete, so suppose $\delta > 0$, and without loss of generality we may take $\delta = 1$. Let $\{x_n\}$ be a minimizing sequence, i.e. $x_n \in K$ and $\|x_n\| = \delta_n \to 1$. Then

$$\frac{1}{2}\left(\frac{x_m}{\delta_m} + \frac{x_n}{\delta_n}\right) = \frac{\delta_n x_m + \delta_m x_n}{\delta_m + \delta_n} \cdot \frac{\delta_m + \delta_n}{2\delta_m \delta_n} = y_{mn} \frac{\delta_m + \delta_n}{2\delta_m \delta_n} ,$$

where, because of the convexity of K, $y_{mn} \in K$, so $\|y_{mn}\| \geqslant 1$. Thus

$$\left\| \frac{1}{2}\left(\frac{x_m}{\delta_m} + \frac{x_n}{\delta_n}\right)\right\| \geq \frac{\delta_m + \delta_n}{2\delta_m \delta_n} \to 1$$

as $m, n \to \infty$. The condition of uniform convexity then implies that $\{\delta_n^{-1} x_n\}$ is a Cauchy sequence, and since $\delta_n \to 1$, $\{x_n\}$ is also a Cauchy sequence, whose limit is then the sought element of K with smallest norm. Unicity follows from Theorem 4.1.1.

◇

<u>4.2 Orthogonality in L^p-spaces</u>. Let (Q, ρ) be a positive measure space, let $X = L^p(\rho) = L^p(Q, \rho)$ and, as before, let Y denote a closed subspace of X. From 4.1 we know that for $1 < p < \infty$, to each g in X, there is a unique g^* best approximating g from Y. We shall now give a characterization of that best approximating element; the case $p = 1$ is somewhat more complicated and will be stated separately.

We recall that f is said to be <u>orthogonal</u> to Y, written $f \perp Y$, if and only if $\|f\| \leq \|f + k\|$ for all $k \in Y$; by considering $f = g - g^*$, the characterization of elements of best approximation is reduced to the study of orthogonality (just as in Chapter 2). The next two theorems characterize orthogonality in L^p spaces, $p < \infty$. (For $p = \infty$ there is no useful analogue; the study of continuous functions on a compact space has been treated in Chapter 2.)

<u>4.2.1 Theorem</u>. <u>Let</u> Y <u>be a closed subspace of</u> $L^p(\rho)$, $1 < p < \infty$. <u>An element</u> $f \in L^p(\rho)$ <u>is orthogonal to</u> Y <u>if and only if</u>

$$\int |f|^{p-1}\, \overline{\operatorname{sgn}}\, f \cdot k \, d\rho = 0 \tag{1}$$

<u>for all</u> $k \in Y$.

<u>Proof</u>. (i) Observe that $g = |f|^{p-1}\,\overline{\operatorname{sgn}}\, f$ is in $L^{p'}$ and $\int |g|^{p'} d\rho = \int |f|^p \, d\rho$ (where p' denotes as usual the conjugate exponent, $p' = p/(p-1)$). Clearly we may assume f is not the zero function, and it is no loss of generality to suppose $\|f\|_p = 1$. Assume that (1) holds. Then for $k \in Y$,

$$\|f\|_p = 1 = \int fg \, d\rho = \int (f + k)g \, d\rho \leq \|f + k\|_p \, \|g\|_{p'} = \|f + k\|_p$$

proving that $f \perp Y$ (observe that this argument is valid also for $p = 1$).

(ii) Suppose now $f \perp Y$, i.e. the distance from f to Y is $\|f\|$. By a standard corollary to the Hahn-Banach theorem there exists a linear functional M on $L^p(\rho)$ such that $Mf = 1$, $MY = 0$ and $\|M\| = \|f\|_p^{-1}$. This M is then representable by some element h of $L^{p'}(\rho)$, $Mf = \int fh \, d\rho$ where $\|h\|_{p'} = \|f\|_p^{-1}$. Therefore $\int fh \, d\rho = \|f\|_p \, \|h\|_{p'}$, and by the conditions for equality to hold in the Hölder inequality we deduce

$$h(x)\ f(x) \geqslant 0\ , \qquad \rho\text{ - a.e.}$$

$$|h(x)|^{p'} = \lambda\ |f(x)|^p\ , \qquad \rho\text{ - a.e.}$$

for a suitable constant λ. Hence $h(x) = \lambda\ |f(x)|^{p-1}\ \overline{\operatorname{sgn}}\ f(x)$, and the condition $MY = 0$ yields (1). ◇

Observe the linearity of the orthogonality condition (1); that is, if f is orthogonal to each of the closed subspaces Y_1 and Y_2, it is also orthogonal to their direct sum. This is of course not true in arbitrary normed linear spaces (e.g. not in C(Q)). Thus in L^p with $1 < p < \infty$, orthogonality of f to Y is equivalent to orthogonality of f to each one-dimensional subspace of Y, i.e. can be characterized in terms of <u>elementwise</u> orthogonality (this relation between a pair of elements is, however, not a symmetric one for $p \neq 2$). For further discussion of orthogonality in normed spaces, and in particular its relation to Gateaux differentiability, see JAMES. The same argument used to establish Theorem 4.2.1 also yields

<u>4.2.2 Theorem.</u> <u>Let</u> Y <u>be a closed subspace of</u> $L^1(\rho)$. <u>An element</u> $f \in L^1(\rho)$ <u>is orthogonal to</u> Y <u>if and only if there is a function</u> φ <u>with</u> $\|\varphi\|_\infty = 1$ <u>such that</u>

$$\int f\varphi \, d\rho = \int |f| \, d\rho$$

<u>and</u>

$$\int \varphi k \, d\rho = 0\ , \qquad \underline{\text{all}}\ k \in Y.$$

<u>In particular, if</u> $\{x\colon f(x) = 0\}$ <u>has measure zero, these conditions are equivalent to</u>

$$(*) \qquad \int \overline{\operatorname{sgn}}\ f \cdot k \, d\rho = 0\ , \qquad \underline{\text{all}}\ k \in Y\ .$$

<u>Remark.</u> For a detailed discussion of best approximation in normed linear spaces (not merely L^p spaces) see SINGER_1.

4.2.3 Exercises. a) Show that (*) does not in general imply $f \perp Y$.

b) Show that the n.a.s.c. in Theorem 4.2.2 is equivalent to the following: for every $k \in Y$,

$$\left|\int_E \overline{\operatorname{sgn} f} \cdot k \, d\rho\right| \leqslant \int_{Q \setminus E} |k| \, d\rho$$

where $E = \{x: f(x) \neq 0\}$. (For further details about orthogonality in L^1, see KRIPKE & RIVLIN.)

c) Show that Theorem 2.6.5 is valid when $C(Q_1 \times Q_2)$ is replaced by $L^p(I \times I)$, $C(Q_1)$ and $C(Q_2)$ by $L^p(I)$, if and only if $p = 2$.

d) Show that the derivative of the Chebyshev polynomial of order $n + 1$ is orthogonal, in $L^1(I)$, to $\mathcal{P}_{n-1}$. Use this to compute the distance, in $L^1(I)$, from x^n to $\mathcal{P}_{n-1}$.

e) The Chebyshev polynomial C_n is orthogonal to $\mathcal{P}_{n-1}$ in the uniform norm, and also in the <u>total variation</u> norm $\|f\| = V(f)$ = total variation of f on I. Is this also the case for L^p norms?

f) If $f^{(n)}(x) > 0$ on $(-1, 1)$, a best L^1 approximation to f from $\mathcal{P}_{n-1}$ is the Lagrange interpolating polynomial which agrees with f at the roots of $C'_{n+1}(x)$, i.e. at the points $x_k = \cos (k\pi/n + 1)$, $k = 1, 2, \ldots n$.

g) The best L^1 approximation to $f \in L^1(I)$ from $\mathcal{P}_n$ need not be unique, but it is when f is continuous.

h) Prove that $\cos (nx + a)$ is orthogonal to $\mathcal{T}_{n-1}$ in every L^p space $(1 \leqslant p \leqslant \infty)$,

(i) by directly demonstrating that 0 is a best approximation

(ii) by verifying that the criteria given in Theorems 4.2.1 and 4.2.2 are satisfied.

(Hint for (i): in case $1 < p \leqslant \infty$ we know by previous general theorems that the b.a. is unique. Show that there <u>exists</u> a b.a. which has period $2\pi/n$, and hence <u>the</u> b.a. has this period, and deduce therefrom that it is 0.)

<u>4.2.4 Functional equations satisfied by a b.a.</u> The last exercise involves a general principle: certain structural properties of a function are "inherited" by at least one of its best approximations from a given set S (assuming that <u>some</u> b.a. exists), provided that S has suitable invariance properties. One general formulation of this is as follows. Let X be a Banach space of functions defined on a set E, and G a group of mappings of E onto itself with the property that whenever $f \in X$ and $g \in G$, $f \circ g$ belongs to X and has the same norm as f ("X is G-invariant"). <u>If</u> S <u>is a convex</u> G-<u>invariant subset of</u> X <u>which contains at least one closest element to some</u> $h \in X$ <u>for which</u> $h \circ g = h$, <u>all</u> $g \in G$, <u>then it contains a closest element</u> k <u>which also satisfies</u> $k \circ g = k$, <u>all</u> $g \in G$.

To prove a rigorous version of this principle some additional assumptions would have to be introduced, which we do not wish to enter into here, but the idea is as follows. (We assume that G is a compact topological group, so that there is a (say, right) Haar measure on it of total mass one, which we denote by dg.) Let p denote a closest element to h from S, and write $d = \|h - p\|$. Then, for each $g \in G$, $d = \|h \circ g - p \circ g\| = \|h - p \circ g\|$. Now, with some reasonable hypotheses, the (X-valued) integration $\int (p \circ g)dg$ makes sense and defines an element $k \in X$. Since each $p \circ g$ is a closest element to h, so is k which is in the convex hull of these. Also, k lies in the convex hull of S (hence in S), and since it is easily verified that k has the desired invariance properties the proposition is established. (In case the group G is <u>finite</u>, the above integral is simply a finite average over the elements of G, and the proof is rigorous as it stands without further assumptions. This is sufficient for the previous exercise, where X is a space of functions defined on the circle and G is a finite subgroup of the group of rotations.)

For a related discussion, along somewhat different lines, see MEINARDUS, p. 24.

<u>4.2.5 Exercises.</u> a) Formulate a rigorous theorem embodying the above principle.

b) In the same framework, suppose further that h has a <u>unique</u> closest element in S. If now h satisfies the equation $h = \int (h \circ g)d\rho$ where ρ is some positive measure on G with total mass one, show that this closest element satisfies the same equation (here G needn't be compact).

c) Let X denote the uniformly continuous functions on R^n, with sup norm, and S a closed translation-invariant subspace. If $f \in X$ has period ξ (i.e. $f(x) = f(x + \xi)$) for some $\xi \in R^n$, and S contains at least one closest element to f, it contains a closest element that has period ξ.

d) With X as in c), let S be a closed rotation-invariant subspace (i.e. invariant with respect to orthogonal transformations of the independent variable.) Prove that if $f \in X$ is radial (i.e. a function of $|x|$), and S contains at least one closest element to f, it contains a closest element that is radial. Prove also the L^p version of this.

e) A function is said to be anti-radial if its integral over every ball centered at the origin is zero. Prove, with X as in c), that an anti-radial function is orthogonal to the subspace of radial functions, and a radial function is orthogonal to the subspace of anti-radial functions. Are these statements valid in $L^p(R^n)$?

f) Let Q denote the unit sphere in R^3, and $X = C(Q)$, S = set of polynomials of degree at most m in x_1, x_2, x_3. Prove, if f is a zonal function (i.e. a function of $x_1^2 + x_2^2$ and x_3) it admits a best approximation in S which is zonal. Can there be more than one b.a.?

g) Let Q denote the unit ball in R^n, $X = C(Q)$ and S the set of polynomials of degree at most m in $x_1, \ldots, x_n$. Prove, if f satisfies the Laplace equation in the interior of Q, it admits a b.a. in S which satisfies the Laplace equation. Can there be more than one b.a.?

We give next an application of L^p orthogonality to analytic functions. Suppose we wish to find the distance from $f \in L^p(T)$ to (the set of boundary values of) the Hardy space H^p (T denotes as usual the unit circle), where $1 < p < \infty$ (cf. DUREN, HOFFMAN). Let f^* denote the element in H^p closest to f and write $F(z) = f(z) - f^*(z)$. Then $F \perp H^p$ and this is equivalent to

$$\int_0^{2\pi} |F(z)|^{p-1} \, \overline{\operatorname{sgn}} \, F(z) \, g(z) \, d\theta = 0 , \qquad z = e^{i\theta}, \quad g \in H^p ,$$

from which we may conclude that $|F(z)|^{p-1}\,\overline{\mathrm{sgn}}\,F(z) = zh(z)$, i.e. $|F(z)|^p = zh(z)F(z)$ a.e. on T for some $h \in H^{p'}$. We can formulate this result thus:

<u>4.2.6 Theorem.</u> <u>Necessary and sufficient condition that</u> $F \in L^p(T)$, $1 < p < \infty$, <u>be orthogonal to</u> H^p <u>is that there exist a function</u> $h \in H^{p'}$ $(p' = p/(p-1))$ <u>such that</u> (with $z = e^{i\theta}$)

$$zh(z)\ F(z) \geqslant 0\ , \tag{2}$$

$$|h(z)| = |F(z)|^{p-1} \tag{3}$$

<u>hold almost everywhere on</u> T. <u>For</u> $p = 1$ <u>the condition is sufficient</u>; <u>it is necessary if</u> F <u>is different from zero a.e.</u>.

Observe that

$$\operatorname{dist}(F, H^p) = \|F\|_{L^p} = \left\{\frac{1}{2\pi}\int_0^{2\pi} |F(e^{i\theta})|^p\, d\theta\right\}^{1/p} = \left\{\frac{1}{2\pi i}\int_{|z|=1} h(z)\ F(z)\ dz\right\}^{1/p}$$

As an application, we shall find the best H^p approximant to $f(z) = (z - a)^{-1}$, $|a| < 1$; with the aid of (2) and (3) very little guesswork is needed. Here $F\ (= f - f^*,\ f^* \in H^p)$ will have a simple pole at $z = a$. Now, $z \in T$, $z/(z-a)(1-\bar{a}z) = |z - a|^{-2} > 0$, so it is natural to look for h and F satisfying

$$h(z)\cdot F(z) = \frac{\text{const.}}{(z-a)(1-\bar{a}z)}\ , \qquad z \in T\ .$$

Since F has a simple pole at $z = a$, let us try

$$F(z) = A\,\frac{(1-\bar{a}z)^\alpha}{z-a}\ , \tag{4}$$

where A and α have to be adjusted so that (2) and (3) are satisfied. From (3)

$$|h(z)| = \left|A\,\frac{(1-\bar{a}z)^\alpha}{z-a}\right|^{p-1}\ , \qquad z \in T\ ,$$

and a holomorphic solution of this is (since $|1 - \bar{a}z| = |z - a|$ on T)

$$h(z) = A^{p-1}(1-\bar{a}z)^{(\alpha-1)(p-1)}. \tag{5}$$

By (4) and (5),

$$zh(z)\ F(z) = A^p z \frac{(1 - \bar{a}z)^{(\alpha-1)(p-1)+\alpha}}{z - a} = A^p \frac{(1 - \bar{a}z)^{\alpha p - p + 2}}{|z - a|^2}$$

on T, so (2) will hold if $A > 0$ and $\alpha p - p + 2 = 0$, i.e. $\alpha = (p - 2)/p$. Since the residue of F at $z = a$ is 1, we must have $A(1 - \bar{a}a)^{\alpha} = 1$, hence $A = (1 - \bar{a}a)^{-(p-2)/p}$. Finally, since on T we have $zh(z)\ F(z) = |F(z)|^p$,

$$\frac{1}{2\pi}\int_0^{2\pi} |F(e^{i\theta})|^p\, d\theta = \frac{1}{2\pi i}\int_T h(z)\ F(z)\ dz = h(a)$$

(by residues), $= (1 - \bar{a}a)^{-(p-1)}$. Summing up all this we have this result:

<u>4.2.7 Theorem</u>. <u>Let</u> $1 < p < \infty$, <u>and</u> $|a| < 1$. <u>Then, there is a unique</u> $f^* \in H^p$ <u>which best approximates</u> $f(z) = (z - a)^{-1}$, <u>namely</u>

$$f^*(z) = \frac{1}{z - a}\left[1 - \left(\frac{1 - \bar{a}z}{1 - \bar{a}a}\right)^{\frac{p-2}{p}}\right]$$

<u>and</u> dist $((z - a)^{-1}, H^p) = (1 - |a|^2)^{-(p-1)/p}$.

<u>Remarks</u>. This result is also valid for $p = 1$ (although <u>uniqueness</u> requires some further checking, since F, h above (in the case $p = 1$) satisfy the conditions sufficient for orthogonality. The correct solution for sup norm approximation ($p = \infty$) (already known to us from Chapter 2) also follows as a limiting case.

Note that dist $((z - a)^{-1}, H^1) = 1$, i.e. is independent of a, and that for $p = 2$, $f^*(z) = 0$, i.e. $(z - a)^{-1} \perp H^2$ (which is easily verified by other methods, since in this case $\perp$-orthogonality coincides with the usual inner-product orthogonality of H^2). For $p = \infty$, $f^*(z) = \bar{a}(1 - |a|^2)^{-1}$ = const. and dist $((z - a)^{-1}, H^{\infty}) = (1 - |a|^2)^{-1}$, in agreement with 2.6.3.2. (Incidentally, the technique of solving a sup norm problem as a limiting case of L^p for $p \to \infty$ is occasionally useful, both in theoretical and in numerical work.)

<u>Exercises</u>. a) Prove the uniqueness of the solution in the case $p = 1$.

b) Prove directly that $(z - a)^{-1} \perp H^2$ (Hint: look at the Fourier expansion of $(e^{i\theta} - a)^{-1}$.)

By means of the derived results we can solve the following problem: Given a, $|a| < 1$, to determine $\sup |g(a)|$; $g \in H^p$, $\|g\|_p = 1$.

Indeed,

$$g(a) = \frac{1}{2\pi i} \int_{|z|=1} g(z)\left(\frac{1}{z - a} - f(z)\right) dz$$

for all $f \in H^{p'}$. Thus $|g(a)| \leq \|g\|_p \|(z - a)^{-1} - f\|_{p'}$, so if f is chosen as f^*, the best approximation to $(z - a)^{-1}$ in $H^{p'}$, we get

$$(1) \qquad |g(a)| \leq \operatorname{dist}\left((z - a)^{-1}, H^{p'}\right) = (1 - |a|^2)^{-(p'-1)/p'} = (1 - |a|^2)^{-1/p} .$$

To see that this bound is best possible, choose for g the function h from the preceding analysis, normalized so that $\|h\|_p = 1$, i.e. take (note that the index that was called p has now become p')

$$g(z) = \text{const.}\ (1 - \bar{a}z)^{-2(p'-1)/p'} = (1 - \bar{a}a)^{1/p}\, (1 - \bar{a}z)^{-2/p}$$

and for this function equality is attained in (1). To summarize:

<u>4.2.8 Theorem</u>. <u>For</u> $g \in H^p$, $1 \leq p \leq \infty$, <u>and</u> $|a| < 1$, <u>the sharp inequality</u>

$$|g(a)| \leq \|g\|_p (1 - |a|^2)^{-1/p}$$

<u>holds, equality being attained if and only if</u> $g(z)$ <u>is a constant multiple of</u> $(1 - \bar{a}z)^{-2/p}$.

<u>Remark</u>. The problems treated in this section are the simplest case of "dual extremal problems" for analytic functions. If we replace $(z - a)^{-1}$ by $(z-a)^{-1}(z-b)^{-1}$ the analysis becomes very much more complicated. For other explicitly solvable cases, as well as further theoretical discussion, see MACINTYRE & ROGOSINSKI, ROGOSINSKI & SHAPIRO, HAVINSON$_1$, DUREN. (In 5.1 below we shall discuss these questions from a slightly different viewpoint.)

<u>4.2.9 A Theorem of Szegö</u>. As another illustration of variational methods in L^p problems, consider the following problem:

<u>Let</u> K <u>be a measurable function on the circle, satisfying</u> $0 < a \leq K(e^{it}) \leq A$;

<u>find the minimum of</u> $I(f) = (2\pi)^{-1} \int_0^{2\pi} |f(e^{it})|^p K(e^{it})\, dt$ <u>over all</u> $f \in H^p$, $f(0) = 1$.

<u>Solution</u>. We restrict ourselves here to the case $1 < p < \infty$. Because of the hypotheses on K, the $L^p(K\,dt)$ norm is equivalent to the $L^p(dt)$ norm, hence $\{f \in H^p: f(0) = 1\}$ is a closed convex set in the (uniformly convex) Banach space $L^p(K\,dt)$, hence has a unique element f_o of least norm. Now, $Y = \{f \in H^p: f(0) = 0\}$ is a closed subspace of $L^p(K\,dt)$, and the minimal property of f_o implies it is orthogonal to Y. Hence

$$\int_0^{2\pi} |f_o(e^{it})|^{p-1}\, \overline{\mathrm{sgn}}\, f_o(e^{it})\, g(e^{it})\, K(e^{it})\, dt = 0\,, \qquad g \in Y$$

This implies that $|f_o(e^{it})|^{p-1}\, \overline{\mathrm{sgn}}\, f_o(e^{it})\, K(e^{it})$ (which is in $L^{p'}(T)$) is (a boundary value) of class $H^{p'}$, hence $|f_o(e^{it})|^p K(e^{it})$ is (a boundary value) of class H^1 and, being real valued, it is constant. Thus $|f_o(e^{it})| = CK(e^{it})^{-1/p}$ a.e. Moreover, f_o has no inner factor, for, if φ were a non-constant inner function dividing f_o, the admissible function $f_1 = \varphi(0)\, \varphi^{-1} f_o$ would have $I(f_1) = |\varphi(0)|\, I(f_o) < I(f_o)$ contradicting the minimal property of f_o.

Therefore, for $|z| < 1$

$$f_o(z) = \exp\left\{-\frac{1}{2\pi}\int_0^{2\pi} \frac{z + e^{it}}{z - e^{it}} \log |f_o(e^{it})|dt\right\} = C \exp\left\{\frac{1}{2\pi p}\int_0^{2\pi} \frac{z + e^{it}}{z - e^{it}} \log K(e^{it})dt\right\}.$$

Letting $z = 0$, we get $1 = f_o(0) = CG(K)^{-1/p}$ where $G(K) = \exp\left\{(2\pi)^{-1} \int \log K\, dt\right\}$ is the geometric mean of K. Thus, $|f_o(e^{it})|^p K(e^{it}) = C^p = G(K)$ a.e., and finally $I(f_o) = G(K)$. <u>Conclusion</u>: <u>The minimum value of</u> $I(f)$ <u>over all</u> $f \in H^p$ <u>with</u> $f(0) = 1$ <u>is the geometric mean of</u> K. <u>The unique minimizing function is that outer function satisfying the normalization at</u> 0 <u>whose modulus on</u> T <u>is a constant multiple of</u> $K(e^{it})^{-1/p}$.

It is interesting to observe that the value of the minimum does not depend on p. From this we can conclude easily that the result is valid also for $p = 1$ (apart possibly from the uniqueness). It is possible, by a rather tedious limiting procedure, to extend the result to an arbitrary non-negative K, thereby obtaining a classical result of Szegö (cf. HELSON p. 18, GRENANDER & SZEGÖ Chapter 3).

Exercises. a) Carry out the details of the case $p = 1$, and investigate the uniqueness.

b) Prove that inf $I(f)$ over all polynomials $f(z) = 1 + a_1 z + \dots a_n z^n$, with n arbitrary, is $G(K)$.

c) Prove that if $G(K) = 0$, the functions $\{e^{int}\}_{n=0}^{\infty}$ span $L^p(Kdt)$.

d) Prove that if $G(K) > 0$, $\{e^{int}\}_{n=0}^{\infty}$ do not span $L^p(K\,dt)$.

e) Assuming K bounded above, the (boundary values of) functions in H^p are a subset of $L^p(K\,dt)$. Determine for which K this set is closed.

By essentially the same method, we can get similar results when K dt is replaced by a measure that is not necessarily absolutely continuous. For simplicity we shall only deal with the case $p = 2$. We shall outline a proof of:

4.2.10 Theorem. Let ρ be a positive bounded measure on T and ρ_a its absolutely continuous part (with respect to the Haar measure dt). If h, the Radon-Nikodym derivative of ρ_a with respect to dt, satisfies $G(h) = 0$, in other words $\int \log h(e^{it})dt = -\infty$, then the functions $\{e^{int}\}_{n=0}^{\infty}$ span $L^2(\rho)$.

Proof. Let us write $\psi(t) = e^{it}$, and assume (to avoid 2π factors) Haar measure normalized so that $\int_T dt = 1$. It is sufficient to prove the following: $\underline{1}$ lies in the closed span S of $\{\psi^n\}_{n=1}^{\infty}$. For then, $\overline{\psi}$ lies in the closed span of $\{\psi^n\}_{n=0}^{\infty}$ (since multiplication by $\overline{\psi}$ is an isometry in $L^2(\rho)$) and by iteration so does $\overline{\psi}^2, \overline{\psi}^3, \dots$. Thus, taking for granted that $\{\psi^n\}_{n=-\infty}^{\infty}$ span $L^2(\rho)$ (which is a consequence of the Weierstrass approximation theorem, to be proved later) we will be done. Suppose, then, that $\underline{1}$ is not in S. Let F be the orthogonal projection from $\underline{1}$ into S. Then, denoting by (,) the inner product in $L^2(\rho)$ we have

(i) $$(\underline{1} - F, \psi^n) = 0, \quad n = 1, 2, \dots$$

(ii) $$(\underline{1} - F, F\psi^n) = 0, \quad n = 1, 2, \dots$$

The latter relation is valid because $F\psi^n$ is in S (S being stable under multiplication by ψ). Hence $(\underline{1} - F, (\underline{1} - F)\psi^n) = 0$, that is $\int |\underline{1} - F|^2 \psi^n \, d\rho = 0$, $n = 1, 2, \ldots$ and taking complex conjugates we see that the latter holds for all $n \neq 0$. We conclude

$$(1) \qquad |1 - F(e^{it})|^2 \, d\rho = c \, dt$$

for some positive constant c. Let now $\rho = \rho_a + \rho_s$ be the decomposition of ρ into its absolutely continuous and singular parts. Then, from (1)

$$|1 - F(e^{it})|^2 \, d\rho_a + |1 - F(e^{it})|^2 \, d\rho_s = c \, dt$$

showing that

$$(2) \qquad |1 - F(e^{it})|^2 \, d\rho_a = c \, dt \; ,$$

$$|1 - F(e^{it})|^2 \, d\rho_s = 0 \; .$$

The latter equation says that $F(e^{it}) = 1$ except on a set of ρ_s-measure zero. Let now $g = (\underline{1} - F)^{-1}$. Then (2) shows $g \in L^2(dt)$.

We claim that, in fact, g is (the boundary value of) a function of class H^2. Indeed, because of (2), for $n \geqslant 1$, $\int g \psi^n \, dt = \int (c\overline{g})^{-1} \psi^n \, d\rho_a = c^{-1} \int (\underline{1} - \overline{F})\psi^n \, d\rho_a = c^{-1} \int (\underline{1} - \overline{F})\psi^n \, d\rho$ (since $\underline{1} - \overline{F}$ vanishes ρ_s - almost everywhere) and this is zero by (1).

We have therefore shown that the Radon-Nikodym derivative $d\rho_a/dt$ equals $|g(t)|^2$, i.e. it is equal to the modulus of (the boundary value of) a function of class H^1. Therefore the desired result is a consequence of the following proposition:

(*) If f is a non-null function in H^1, then $\int \log |f(e^{it})| \, dt > -\infty$.

Indeed, assuming (*) we have an immediate contradiction (to the assumption that $\underline{1}$ is not in S) and the Theorem is proved.

As for (*), it is a standard elementary result in the theory of H^p spaces, and shall not be proven here (it is of course essential that (*) can be proven independently of Theorem 4.2.10; in some treatments, e.g. HOFFMAN, the order is reversed and (*) is deduced from Theorem 4.2.10, for which an independent proof is given. This is perhaps the most elegant approach, insofar as the approximation theorem is proved by "pure" approximation theory, without recourse to properties of analytic functions, and this approach is essential when one wishes to prove (*) for various classes of generalized H^1 spaces (cf. HOFFMAN); it would take us too far afield to delve into this fascinating subject here.)

<u>Exercises</u>. a) Prove that if ρ is singular, $\{e^{int}\}_{n=0}^{\infty}$ span $L^2(\rho)$.

b) Prove the converse of Theorem 4.2.10: <u>if</u> $\int \log (d\rho_a/dt)\, dt > -\infty$, $\{e^{int}\}_{n=0}^{\infty}$ <u>fail to span</u> $L^2(\rho)$.

c) Prove (*) (hints: treat first the case $f \in H^{\infty}$, using Jensen's theorem together with Fatou's lemma; for the general case, show that an H^1 function is the quotient of two bounded ones).

d) Carry out the above analysis in $L^p(\rho)$, $1 < p < \infty$ and deduce the results for $p = 1$ by a limiting argument.

e) Prove that if T_o is a proper subarc of T the functions $\{e^{int}\}_{n=0}^{\infty}$ span $C(T_o)$ (in the uniform norm).

f) For which non-negative continuous functions K on the circle is it true that for every $f \in C(T)$, Kf is uniformly approximable on T by finite linear combinations of $\{K\, e^{int}\}_{n=0}^{\infty}$?

g) When ρ is such a measure that the closed span J of $\{e^{int}\}_{n=0}^{\infty}$ in $L^2(\rho)$ is proper, give an "explicit" description of the elements of J.

h) Prove that if a bounded measure ρ on T satisfies $\int e^{int}\, d\rho = 0$, $n = 1, 2, \ldots$, these equations hold also when ρ is replaced by its singular part ρ_s, and hence that $\rho_s = 0$ (Theorem of F. and M. Riesz).

<u>Remarks</u>. Theorem 4.2.10 and its converse comprise (a generalization by Kolmogorov and Krein of) a theorem of Szegö. The idea to deduce this theorem by a variational argument is due to Helson and Lowdenslager. The subject is intertwined closely with, among many other things, the F. and M. Riesz theorem on "analytic measures", singly and doubly invariant subspaces of the shift operator, prediction theory, Toeplitz operators, the representation theory of H^p spaces (classical and generalized), and orthogonal polynomials (which was Szegö's own point of departure). The reader wishing

to pursue the study of these matters could consult the books of HOFFMAN, HELSON, and GRENANDER & SZEGÖ, in which of course further references may be found.

Chapter 5. Applications of the Hahn-Banach Theorem and Dual Extremal Problems

5.1 Dual Extremal Problems. We have already encountered in Chapter 2 the situation where a minimum problem was intimately related to a certain "dual" maximum problem. This is actually a general Banach space phenomenon, not restricted merely to sup norm problems as we shall see. Let us first look at a simple numerical example. Suppose we wish to find a measure σ on $J = [0, 1]$ of smallest total variation such that the "moment" conditions $\int d\sigma = 1$, $\int x \, d\sigma = 2$ are satisfied.

We have, for any "competing" σ, and arbitrary real a, b

$$|2a + b| = \left|\int (ax + b) \, d\sigma\right| \leq V(\sigma) \max_{x \in J} |ax + b| .$$

In order to obtain here the best possible lower bound for $V(\sigma)$, we are thus led (observe the homogeneity with respect to a, b) to minimize $\max_{x \in J} |ax + b|$ subject to the constraint $2a + b = 1$, i.e. to compute

$$\lambda = \min_{a \in R} \max_{x \in J} |ax + (1 - 2a)| = \min_{a \in R} \max (|1 - 2a|, |1 - a|)$$

and simple graphical considerations show that $\lambda = 1/3$, the minimum being attained for the unique choice $a = 2/3$. We thus get the lower bound $V(\sigma) \geq 3$. In order for equality to be attained, we must find a "competing" σ such that $\left|\int p \, d\sigma\right| = V(\sigma) \max_{x \in J} |p(x)|$ where $p(x) = (2/3)x - (1/3)$. Thus, σ must be supported on the subset of J where $|p(x)|$ attains its maximum, i.e. the set $\{0, 1\}$, and have masses of opposite signs at those points. Since the only measure supported on the set $\{0, 1\}$ satisfying the moment conditions (namely, that which gives "masses" - 1, 2 to the points $x = 0$, $x = 1$ respectively) does satisfy the latter condition, this measure solves the original minimum problem, and it is the unique solution.

Observe that what we did was to pass from the original minimum problem in the space of measures (the dual space of C(J)) to a different ("dual") extremal problem (whether we choose to call it a "minimum" or "maximum" problem depends of course on how we fix our normalization) in C(J). The latter problem could be solved by inspection, and this solution forced a lower bound for $V(\sigma)$. What is remarkable, and appeared as a "coincidence", is that this lower bound is indeed the exact minimum in the original problem. Moreover, the solution of the dual problem enabled us to find all the extremals σ. We have already seen other examples of the same sort in Chapter 2, and there developed a duality theory sufficient to explain the "resonance" between an extremal problem and its dual, so that in particular we know a priori in the above example that λ^{-1} will be the infimum of $V(\sigma)$ over all admissible σ. Let us now consider the analogous problem in another norm: we seek a bounded measurable function f on J such that $\int f(x)\,dx = 1$, $\int x\,f(x)\,dx = 2$ and the essential supremum $\|f\|_\infty$ is minimum. We have

$$|2a + b| = \left|\int (ax + b)\,f(x)\,dx\right| \leqslant \|f\|_\infty \int |ax + b|\,dx$$

hence, for any competing function f, $\|f\|_\infty \geqslant \lambda^{-1}$, where now

$$\lambda = \min_{a \in R} \int_0^1 |ax + (1 - 2a)|\,dx$$

Since

$$\int_0^1 |ax + (1 - 2a)|\,dx = \begin{cases} (5a^2 - 6a + 2)/2a\,, & \text{if } \frac{1}{2} \leqslant a \leqslant 1 \\ |1 - (3/2)a|\,, & \text{otherwise} \end{cases}$$

we compute that the minimum is attained for $a = (2/5)^{1/2}$; hence $\lambda = (10)^{1/2} - 3$ and so $\|f\|_\infty \geqslant (10)^{1/2} + 3$. For the critical choice of a, the function $ax + (1 - 2a)$ changes sign at $x^* = (1/2)(4 - (10)^{1/2})$. Therefore, if $(10)^{1/2} + 3 = \lambda^{-1}$ is to be an attained value of $\|f\|_\infty$, the extremal f must be a two-step function taking the values $\pm\lambda^{-1}$, and changing sign at $x = x^*$. In fact, it is readily verified that the function

$$f(x) = \begin{cases} -\lambda^{-1}\,, & 0 \leqslant x \leqslant x^* \\ \lambda^{-1}\,, & x^* < x \leqslant 1 \end{cases}$$

does satisfy the moment conditions, and hence is the unique extremal function. Once again, thanks to a "happy coincidence", the method of duality has led to a solution.

Exercises. a) Solve the analogous problem in L^p, where $1 < p < \infty$.

b) Show that for $p = 1$ no extremal function exists, but inf $\|f\|_1$ over all competing functions is 3.

<u>5.1.1 The abstract basis of duality</u>. We now wish to clarify, in general terms, what "dual extremal problems" are, and why the method used above worked. We may say at once that at the root of the matter is the Hahn-Banach theorem, in other words "convexity" - specifically, the principle of the separating hyperplane, of which the Hahn-Banach theorem is an abstract version. (Recall that the results of Chapter 2, in particular the duality theorem 2.3.7, were all based on the principle of the separating hyperplane.) One abstract version of "dual extremal principle" is the following proposition.

(Note: there is probably no one proposition which could lay claim to being "the" abstract formulation of the duality in question; rather there is a certain body of results of this nature in functional analysis and convexity theory, of varying degrees of generality, often tailored to special applications such as linear programming, control theory, or (as in our case) function theory, approximation and moment problems.)

<u>Proposition</u>. <u>Let</u> X <u>be a normed linear space and</u> Y <u>a closed subspace of</u> X. <u>Let</u> X^*, Y^* <u>be the dual spaces of</u> X, Y <u>and let</u> Y^o <u>be the set of all elements of</u> X^* <u>that vanish on</u> Y ("<u>annihilator</u>" <u>of</u> Y). <u>Then the following spaces are isometrically isomorphic</u>:

a) $$X^* / Y^o \quad \underline{\text{and}} \quad Y^*$$

b) $$(X / Y)^* \quad \underline{\text{and}} \quad Y^o .$$

It is understood here that the isomorphisms are the "natural" ones, and that the norm in a quotient space is the usual one, i.e. the norm of a coset is the infimum of the norms of its elements. Part a) expresses that the linear functionals on Y are the restrictions of linear functionals on X, whereby two of the latter are identified if they agree on Y. Part b) expresses that those linear functionals on X which vanish on the subspace Y can be identified with linear functionals on the cosets of Y.

Actually, the above proposition is more convenient to apply when reformulated a bit more explicitly.

5.1.2 Theorem. (Duality Principle). With the same notations as in 5.1.1, for given $x_o \in X$ and $f_o \in X^*$:

a′)
$$\min_{f \in Y^o} \|f_o - f\| = \sup_{\substack{y \in Y \\ \|y\| = 1}} |f_o(y)|$$

b′)
$$\inf_{y \in Y} \|x_o - y\| = \max_{\substack{f \in Y^o \\ \|f\| = 1}} |f(x_o)| .$$

Here min (max) denotes an attained minimum (maximum).

We shall not prove the theorem (a simple deduction from the Hahn-Banach theorem) as it is found in textbooks on functional analysis, but wish to make some additional comments.

Remarks. 1) The theorem can obviously also be formulated as

a′′)
$$\min_{f \in Y^o} \|f_o - f\| = (\inf_{\substack{y \in Y \\ f_o(y)=1}} \|y\|)^{-1}$$

b′′)
$$\inf_{y \in Y} \|x_o - y\| = (\min_{\substack{f \in Y^o \\ f(x_o)=1}} \|f\|)^{-1} .$$

2) The theorem extends to the situation where the subspaces Y, Y^o of X, X^* resp. are replaced by convex sets K, K^* containing the origins of X, X^* resp.; for the appropriate formulation, and a full discussion, see DEUTSCH & MASERICK, SINGER$_1$, IOFFE & TIHOMIROV.

3) It is interesting to observe that there is an attained minimum in a′) and an attained maximum in b′). (The sup in a′) and the inf in b′) are not in general attained, as simple examples show. Concerning this, cf. BONSALL.) Thus, an interesting consequence of a′) is: if f_o is an element of a dual Banach space X^*, it has a closest element in any subspace of X^* which is the annihilator of some subspace in X. This is not true for arbitrary closed subspaces of X^* (exercise: construct a counter-example). A consequence of b′) is that on the unit sphere of a subspace W of X^* which

is the annihilator of some subspace in X, each linear functional of the special type that arises from an element of X attains a maximum. What is really relevant in the latter instance is not that W is an annihilator, but (what is equivalent to its being an annihilator) that it is weak* closed, and therefore its unit sphere is a compact subset of X^* when the latter is endowed with its weak* topology. Since, for each $x_0 \in X$, the function $f \to |f(x_0)|$ is continuous on X^* for the weak* topology (by the very definition of this topology), it attains a maximum on the unit sphere of W. We have, in sum, a useful principle regarding the attainment of extrema:

(i) Every element in a dual Banach space X^* has a closest element in each weak* closed subspace of X^*.

(ii) For each $x_0 \in X$, $|f(x_0)|$ attains a maximum as f ranges over the unit sphere of a weak* closed subspace of X^*.

5.1.3 Exercises. a) Prove Theorem 5.1.2.

b) Prove that a subspace of X^* is the annihilator of some closed subspace of X if and only if it is weak* closed.

c) Give examples to show that the closest element in assertion (i) above, and the maximizing element in (ii), need not be unique; also examples to show that "weak* closed" cannot be replaced by "closed".

d) Let H^p (where $1 \leqslant p \leqslant \infty$) denote the (boundary values of) the Hardy class H^p for the disc (HOFFMAN, DUREN). H^p is a closed subspace of $L^p = L^p(T)$. Prove that H^p is weak* closed in L^p, if $1 < p \leqslant \infty$.

e) Every $f \in L^\infty$ admits a best approximation from H^∞.

(Remark. This best approximation is in general not unique (ROGOSINSKI & SHAPIRO). It is an unsolved problem whether, when f is continuous, its best approximation from H^∞ is also continuous. Recently L. Carleson has shown (unpublished) that this is the case if f is Hölder continuous.)

f) Every $f \in L^1$ admits a best approximation from H^1. (Hint: although L^1 is not a dual Banach space, it can be embedded in a larger space that is, and within which H^1 is a weak* closed subspace.)

Remark. Proposition (i) above, and the results in exercises e) and f) based upon it, are particularly interesting, because they deal with situations where a best approximation exists that is not provided for by the usual criteria (finite dimensionality, uniform convexity). A slightly more general criterion than uniform convexity is reflexivity: within a reflexive Banach space a point always has a closest element in each closed subspace (cf. KÖTHE, p. 343). Neither L^1 nor L^∞ are reflexive, and in fact, each of these spaces possesses closed subspaces and points without closest elements in them (construct examples!).

Exercises (continued). g) Construct an infinite-dimensional weak* closed subspace of $L^\infty(T)$ consisting of continuous functions.

h) Show that the subspace S of $L^\infty(T)$ consisting of functions $\sum_0^\infty a_n e^{i2^n t}$ with $\sum |a_n| < \infty$ is weak* closed; hence every $f \in L^\infty(T)$ admits a best approximation from S.

5.1.4 Let us now see how the preceding discussion explains the results of the examples we considered in the beginning of this section. Suppose, in a Banach space X, we are given linearly independent elements $x_1, \ldots x_n$ and complex numbers $a_1, \ldots \ldots a_n$, not all zero, and consider the set $S \subset X^*$ of continuous linear functionals f which satisfy the interpolation $f(x_i) = a_i$, $i = 1, 2, \ldots n$. (Since S is bounded and weak* closed, we know it contains an element of least norm.) Let Y denote the space spanned by $x_1, \ldots, x_n$, and let f_0 be a fixed element of S.
According to part a') of 5.1.2, there exists, among the functionals which vanish on the x_i, some f_1 for which $\|f_0 - f_1\|$ is minimal, i.e. there exists an $f_2 \in S$ of minimal norm. The value of this norm is, by a'') above, λ^{-1} where $\lambda = \inf \|\sum_{i=1}^n c_i x_i\|$, over $\{c_i\}$ with $\sum_{i=1}^n a_i c_i = 1$ precisely as we found "empirically" in the two examples studied (here the inf is, of course, an attained minimum because Y is finite dimensional). The result thus obtained was found, in the context of certain concrete spaces, by F. Riesz more than a half-century ago, in his study of moment problems.

One of the most fruitful areas of application of duality is to analytic functions. Strangely, these applications seem not to have occurred to anyone for nearly twenty years after the appearance, in BANACH's book, of the necessary tools. An account of the discovery of dual extremal problems in function theory (the pre-Hahn-Banach phase) is in MACINTYRE & ROGOSINSKI, and it is striking how difficult progress was with the earlier ad hoc methods. First Havinson, and later Rogosinski and the author, all working independently, made the observation that the basic result of the whole theory, the equality of a certain sup and a certain inf, was an immediate consequence of the Hahn-Banach theorem (for details see the survey article of $\mathrm{HAVINSON}_1$, also ROGOSINSKI & SHAPIRO, and DUREN; also of interest is LAX). Without wishing here to go deeply into these matters, we discuss two examples, the first the "ancestral" duality phenomenon discovered by Landau, the second an interpolation problem of recent date that has attracted some interest.

5.1.4.1 The Landau problem was: if f is analytic and bounded by one in the unit disc, how large can $\max_{|z|=1} |s_n(z, f)|$ be, $s_n(z, f)$ denoting the partial sum of order n, evaluated at z, of the Taylor series of f? From general Fourier considerations (see 8.6) we get as an upper bound the Lebesgue constant $L_n \sim (4/\pi^2) \log n$, but this is not even asymptotically correct. Landau found the exact upper bound, and the extremal function, as follows. Without loss of generality, we can study the problem of maximizing $|s_n(1, f)|$ on the unit sphere of H^∞. Now,

$$s_n(1, f) = (2\pi)^{-1} \int_0^{2\pi} f(e^{it})(e^{-int} + \dots e^{-it} + 1)\, dt$$

$$= (2\pi)^{-1} \int_0^{2\pi} f(e^{it})(e^{-int} + \dots e^{-it} + 1 + e^{it}\, g(e^{it}))\, dt$$

where g is an arbitrary element of H^1. We get, therefore, if $\|f\|_\infty = 1$,

$$|s_n(1, f)| \leq (2\pi)^{-1} \int_0^{2\pi} |e^{-i(n+1)t} + \dots e^{-it} + g(e^{it})|\, dt .$$

Landau's fundamental idea was to introduce the term involving g before taking absolute values. This gives us the freedom, in the last inequality, to choose g so as to minimize the right hand side, just as we have done earlier to prove 4.2.8. We get thus $|s_n(1, f)| \leq \text{dist}(G, H^1)$ where $G(t) = e^{-i(n+1)t} + \ldots e^{-it}$, so the original problem has now been transformed into finding the best approximation g^* from H^1 to the quite special function G. This Landau achieved by an inspired guess, and with the concrete choice of g^* in hand could easily construct a corresponding $f^* \in H^\infty$ of norm 1 such that $s_n(1, f^*) = \|G + g^*\|_1$, as well as compute explicitly the latter number (for details, see LANDAU). As always occurred in the early study of problems of this kind, the fact that the method led to a best possible result seemed miraculous. In view of the above duality theory, we can look upon the matter as follows. Introduce the pairing

$$(\varphi, F) = \frac{1}{2\pi} \int_0^{2\pi} \varphi(e^{it}) F(e^{it}) e^{it} dt$$

for $\varphi \in L^\infty$, $F \in L^1$. Then, L^∞ is the dual space of L^1, and its subspace H^∞ is the annihilator of H^1. Therefore, by part b') of Theorem 5.1.2, the distance from the above defined element G to H^1 is the attained maximum, on the unit sphere of H^∞, of $|(f, G)| = |s_n(1, f)|$. Actually, although 5.1.2 does not give this directly, H^1 contains a closest element g^* to G, because H^1 can be embedded isometrically as a weak* closed subspace of the space of bounded measures on T (which is a dual space, the dual of C(T)!) From this information it is not hard to gain some further ground, e.g. show that any extremal $f^* \in H^\infty$ must have constant modulus a.e. on T, and even that it is a rational function, and that any g^* is a polynomial of degree n - 1 at most, which is related to every possible f^* in a certain precise way. Unfortunately, this is as far as the general theory can take us! It shows that Landau's method must in principle work, and gives certain guidelines as to how to guess g^*, but provides no algorithm for doing so. Thus, in the present instance, the abstract duality considerations are "icing on the cake", coming after the solution of the problem to explain the reason for the success. But no more can fairly be asked of a general method, and in fact Landau's explicit solution is only possible because of the accidental fact that the partial sums of $(1 - z)^{-1/2}$ have no roots of modulus less than one; indeed, if the functional $s_n(1, f)$ is changed to the more general $\lambda_o f(0) +$

$\lambda_1 f'(0) + \dots \lambda_n f^{(n)}(0)$, an explicit solution of the corresponding maximum problem is not known, and it is extremely unlikely that one can be found. (Theorem 5.1.2 provides us, however, with qualitative information in this case as well as in the preceding.) In the interpolation problem to be discussed next, the situation is different; there the result was discovered with the aid of 5.1.2. There, however, it is a case of an abstract existence proof, the sought object not being constructed explicitly. As a general comment we may remark that, when dealing with extremal problems in analysis, if an explicit solution can be guessed correctly, it is often possible by quite elementary reasoning to verify its extremal property (e.g. having once laid hands on the Chebyshev polynomial, it is very easy to verify its minimax property). This, of course, does not detract from the interest of a general theory, even though in some individual cases the "method" of divine inspiration may give better results.

Exercises. a) Show that for $n = 1$ the maximum of $|s_n(1, f)|$ is $5/4$, the extremals f^* and g^* are unique, and find them.

b) Let $P(z) = a_0 + \dots a_n z^n$, $Q(z) = \overline{a}_n + \dots + \overline{a}_0 z^n = \overline{P(z)} z^n$, $z \in T$. Show that, for $z \in T$, $\operatorname{sgn} P(z)^2 = z^n P(z)/Q(z)$, and hence $R = P^2$ satisfies $\int_0^{2\pi} \overline{\operatorname{sgn}} R(e^{it}) e^{ikt} dt = 0$ for $k \geqslant n + 1$. Use this to find the best approximation from H^1 to the above function G, and hence to solve Landau's problem.

5.2 Universal Interpolating Sequences for H^∞.

(This section involves a fair amount of complex analysis and may be skipped by a reader with other interests.) We study here the problem of characterizing those sequences $\{z_i\}$, $i = 1, 2, \dots$, in the open unit disc D which are such that for each bounded sequence of complex numbers w_i, $i = 1, 2, \dots$, there exists a function f, bounded and analytic in D (i.e. $f \in H^\infty(D)$), satisfying the interpolatory condition $f(z_i) = w_i$, $i = 1, 2, \dots$.

Necessary conditions for such universal interpolating sequences are easily derived. For, then the map $T\colon H^\infty/I \to \ell^\infty$, where $I = \{f \in H^\infty\colon f(z_i) = 0,\ i = 1, 2, \dots\}$, defined by $Tf = (f(z_1), f(z_2), \dots)$ is, by definition, onto, and since it is (because of the quotient construction) bijective, its inverse is bounded, by a theorem of Banach. Thus, if $\{z_n\}$ is any u.i.s., there is a constant M such that whenever $|w_n| \leq 1$, there exists $f \in H^\infty$ satisfying $\|f\|_\infty \leq M$ and $f(z_n) = w_n$, $n = 1, 2, \ldots$. In particular, for each n there is a function f_n in $H^\infty(D)$ with $f_n(z_n) = 1$, $f_n(z_i) = 0$, $i \neq n$ and $\|f_n\|_\infty \leq M$. Hence, by the standard representation theory for H^p spaces, $f_n = B_n g_n$ where $B_n(z) = \prod_{i \neq n} (z - z_i)(1 - \bar{z}_i z)^{-1}(- \operatorname{sgn} \bar{z}_i)$, and $g_n \in H^\infty(D)$ satisfies $\|g_n\|_\infty = \|f_n\|_\infty \leq M$. Now, from $1 \leq |f_n(z_n)| \leq |B_n(z_n)|\ |g_n(z_n)| \leq M\ |B_n(z_n)|$, we get the <u>necessary condition</u>: <u>the numbers</u>

$$(1) \qquad \delta_n \equiv |B_n(z_n)| = \prod_{i \neq n} |z_n - z_i|/|1 - \bar{z}_i z_n| , \qquad n = 1, 2, \ldots$$

<u>are bounded below, whenever</u> $\{z_n\}$ <u>is a u.i.s.</u>.

L. Carleson showed that this condition is also sufficient. Here we shall only outline the first part of the proof, in order to show how, by means of duality, the original problem is transformed into proving a certain inequality for functions of class H^1. We assume henceforth $\delta_n \geq \delta > 0$.

Let p_m be any polynomial (say, for definiteness, the Lagrange interpolating polynomial) satisfying

$$(2) \qquad p_m(z_i) = w_i , \qquad i = 1, \ldots, m,$$

where w_i, $i = 1, 2, \ldots$, are complex numbers bounded by one and let β_m be the (finite) Blaschke product formed with zeroes at $z_1, \ldots, z_m$. Then, for any $g \in H^\infty(D)$, the function $f = p_m + \beta_m g$ belongs to $H^\infty(D)$ and satisfies (2). Since $|\beta_m(e^{i\theta})| = 1$, $\|f\|_\infty$ is the norm in $L^\infty(T)$ of $p_m \bar{\beta}_m + g$. Suppose now that by suitable choice of $g = g_m$, we can make the norm of $f = f_m$ less than a fixed constant A, <u>not depending on</u> m. We will then have solved the required interpolation problem at the first m points, with a fixed bound on the norm, and a standard "normal families" argument then yields a bounded solution of the original problem. Therefore it is sufficient to show that

$$\operatorname{dist} (\bar{\beta}_m p, H^\infty) = \inf_{g \in H^\infty} \|\bar{\beta}_m p + g\|_\infty \leq C ,$$

with C independent of m and $\{w_i\}$ (satisfying $|w_i| \leq 1$). As in the previous section, we set up a pairing between L^∞ and L^1 such that H^∞ becomes the annihilator of H^1. By the Theorem 5.1.2, formula a′), we get

$$\operatorname{dist}(p_m \bar{\beta}_m, H^\infty) = \sup_{\substack{h \in H^1 \\ \|h\|_1 = 1}} \left| \frac{1}{2\pi i} \int_T \frac{h(z)\, p_m(z)}{\beta_m(z)}\, dz \right| .$$

By a residue calculation the integral is

$$\sum_{i=1}^{m} \frac{h(z_i)\, p_m(z_i)}{\beta_m'(z_i)} = \sum_{i=1}^{m} \frac{h(z_i) w_i}{\beta_m'(z_i)} .$$

Now,

$$|\beta_m'(z_i)| = \frac{1}{1 - |z_i|^2} \prod_{\substack{\nu = 1 \\ \nu \neq i}}^{m} \left| \frac{z_i - z_\nu}{1 - \bar{z}_\nu z_i} \right| \geq \delta (1 - |z_i|^2)^{-1} ,$$

so finally (recall that $|w_i| \leq 1$)

$$\operatorname{dist}(p_m \bar{\beta}_m, H^\infty) \leq \sup_{\substack{h \in H^1 \\ \|h\|_1 = 1}} \left| \sum_{i=1}^{m} (1 - |z_i|^2)\, h(z_i) \right| \delta^{-1} .$$

<u>Thus it is sufficient to show that</u>

$$\sum_{i=1}^{\infty} (1 - |z_i|^2)\, |h(z_i)| \leq K(\delta) \cdot \|h\|_1$$

<u>for all</u> h <u>in</u> $H^1(D)$, <u>under the assumption</u> $\delta_n \geq \delta > 0$.

This completes the first part of the proof-transformation of the problem to that of proving a certain inequality for functions of class H^1. The second - rather harder - part of the proof can be done in either of two ways. One variant, due to Shields and the author (for all references, see HOFFMAN), transforms the H^1 problem to an H^2 problem, then by another duality transforms this <u>back</u> to an interpolation problem, but now in H^2, namely, to show that $\{z_n\}$ is a u.i.s. for the space H^2, in the sense of 6.2.8.2 below; finally, the latter problem is solved with the aid of quadratic forms. This proof will be found quite instructive, illustrating several of the techniques we have discussed in the last two sections.

The reader wishing to delve further into dual extremal problems can consult DEUTSCH & MASERICK and $SINGER_1$ (where many other relevant references may be found) for the general theory. See also KARLIN & STUDDEN and works on moment problems referred to there. Further applications to complex analysis may be found in DUREN & WILLIAMS, NEWMAN & $SHAPIRO_3$ and $HAVINSON_2$. For an application of duality to boundary behaviour of analytic functions see $SHAPIRO_7$.

For general discussion of duality in approximation theory and function theory see also $BUCK_{1,2}$. For dual aspects of convex sets and functions see ROCKAFELLAR, also STOER & WITZGALL, where applications to linear programming and optimization are given. Certain topics in control theory are closely related methodologically to the material of this Chapter, see e.g. HERMES & LaSALLE, LUENBERGER.

Exercises. a) We denote by H^p $(1 \leq p \leq \infty)$ the (boundary values of the) Hardy spaces, embedded in the usual way in $L^p(T)$, and place the elements of L^p, L^q (where $q = p/p - 1$) in duality via the bilinear form

$$(f, g) = (2\pi)^{-1} \int_0^{2\pi} f(e^{it})\, g(e^{it})\, e^{it}\, dt\,, \qquad f \in L^p, \quad g \in L^q\,.$$

Prove: For $k \in L^q$, we have

$$\sup_{\substack{f \in H^p \\ \|f\|_p = 1}} |(f, k)| = \operatorname{dist}(k, H^q) = \inf_{g \in H^q} \|k - g\|_q$$

and discuss the existence of a maximizing element $f^* \in H^p$, and of a minimizing element $g^* \in H^q$. What further relationships can you deduce about f^*, g^*? Treat in detail the cases $k(t) = (e^{it} - a)^{-1}$, $|a| < 1$ and $k(t) = e^{-it} + e^{-2it}$, and settle the uniqueness questions for these cases.

b) Carry out a similar analysis, with H^p replaced by $\left\{f \in L^p(R) : \hat{f} \text{ is supported in } (-\infty, 0]\right\}$, and the duality via $(f, g) = \int_{-\infty}^{\infty} f(x)\, g(x)\, dx$. (Here $\hat{f}$ is the Fourier transform, and understood in the distributional sense if $p > 2$; the classes in question can be interpreted as (boundary functions of the) analogs of the Hardy spaces for the upper half-plane.) Treat in detail the cases when $k(t) = (t-a)^{-1}$,

Im a > 0 and k(t) = characteristic function of an interval.

c) Define $E_{p,\tau} = \left\{f \in L^p(R) : \hat{f} \text{ is supported in } [-\tau, \tau]\right\}$ (this space can be identified with (the restrictions to the real axis of) the entire functions of exponential type τ at most, which belong to $L^p(R)$). Give a representation for the dual space of $E_{p,\tau}$, and set up the appropriate "max-min duality" relation corresponding to those in the preceding exercises. Use this to relate the "Bernstein inequality" for $f'(0)$ with a problem of best approximation in the dual space. (For related discussion in the case $p = \infty$, see Chapter 7. LOGAN has detailed discussion of a number of cases, inspired by communication theory, and having also considerable intrinsic interest. Unfortunately, this document, and some other works of Logan where the principles of max-min duality are employed with great virtuousity, remain unpublished at the present moment.)

Chapter 6. Approximation Theory and Extremal Problems in Hilbert Spaces

6.1 Introduction. If we have a Hilbert space H (whose inner product we shall denote by (,)) then "orthogonality" of an element x to a closed subspace Y in the previously introduced normed spaces sense is readily seen to be equivalent to the orthogonality (in the usual Hilbert space sense) of x to each element of Y. The unique closest element to x in Y turns out to be the orthogonal projection of x on Y, and as is well known, the map from x to its orthogonal projection on Y is a linear operation. We assume the reader is familiar with the elementary geometry of Hilbert spaces, and notably with the procedure of computing the orthogonal projection of x onto a subspace Y with the aid of an orthonormal basis for Y. An important result is the formula for the distance d from x to Y, when Y is the n-dimensional space spanned by the linearly independent elements $y_1, \dots y_n$. One has namely

$$d^2 = G(x, y_1, y_2, \dots, y_n)/G(y_1, y_2, \dots, y_n)$$

where $G(x_1, \dots, x_m)$, the Gram determinant (or Gramian) of a collection of vectors $x_1, \dots x_m$ is defined to be the determinant whose entry in the (i, j) position is (x_i, x_j) (cf. DAVIS).

6.1.1 Example. Consider the Hardy space H^2 of the unit disc, and for $|\zeta| < 1$ define $k_\zeta(z) = (1 - \bar{\zeta} z)^{-1}$. Let us compute the distance d from $k_o \equiv 1$ to the span of $\{k_{\zeta_1}, \dots, k_{\zeta_n}\}$. Since $(k_\alpha, k_\beta) = (1 - \bar{\alpha}\beta)^{-1}$ we get

$$d^2 = \frac{\det \begin{vmatrix} 1 & 1 & \dots & 1 \\ 1 & a_{11} & \dots & a_{1n} \\ \vdots & \vdots & & \vdots \\ 1 & a_{n1} & \dots & a_{nn} \end{vmatrix}}{\det \begin{vmatrix} a_{11} & \dots & a_{1n} \\ \vdots & & \vdots \\ a_{n1} & \dots & a_{nn} \end{vmatrix}} = \frac{N}{D}$$

where $a_{ij} = (1 - \bar{\zeta}_i \zeta_j)^{-1}$. To evaluate N, subtract the first column from the succeeding columns. Since $a_{ij} - 1 = \bar{\zeta}_i \zeta_j/(1 - \bar{\zeta}_i \zeta_j)$, we see immediately that $N = \zeta_1 \dots \zeta_n \bar{\zeta}_1 \dots \bar{\zeta}_n D$, and so $d = \prod_{i=1}^{n} |\zeta_i|$.

6.1.2 Exercises. a) Deduce from the above that a bounded analytic function in the unit disc that vanishes at $z = 1 - n^{-1}$ $(n = 1, 2, \dots)$ vanishes identically.

b) Compute the distance from k_ζ to the span of $\{k_{\zeta_1}, \dots k_{\zeta_n}\}$.

c) Compute the distance in $L^2(0, \infty)$ from $e^{-\lambda t}$ to span $\{e^{-\lambda_1 t}, \dots, e^{-\lambda_n t}\}$ where λ, λ_i are any complex numbers whose real parts are positive.

6.2 Hilbert Spaces with Reproducing Kernel

6.2.1 Hilbert function spaces. As we have remarked, the unique closest element to x in a closed subspace Y is the orthogonal projection of x on Y, i.e. the unique $y^* \in Y$ such that $x - y^*$ lies in Y^o, the orthogonal complement of Y. Denoting by P_Y the operator which orthogonally projects elements on Y, i.e. $P_Y x = y^*$, we can say that the map taking each element x to its best approximation from Y is the linear operator P_Y. This linearity is of course special for Hilbert spaces, and leads (at least theoretically) to great simplifications in the study of best approximation, since P_Y can be expressed in an explicit form with the aid of an orthonormal basis for Y. The situation can be rendered still more concrete, and a "formula" given for the orthogonal projection that is often useful in practice, if H is a Hilbert function space, by which we mean a Hilbert space whose elements are complex-valued functions on some set Q and such that the evaluation functionals are bounded. This being so, for each $u \in Q$ there exists (by a theorem of F. Riesz) a unique element $k_u \in H$ such that

$$(f, k_u) = f(u) , \qquad \text{all } f \in H .$$

This k_u is called the reproducing element for the point u. The totality of reproducing elements is the reproducing kernel (r.k.) of H, or, what is equivalent, the r.k. of H is the map

$$(t, u) \to k_u(t)$$

from $Q \times Q$ to the complex numbers. We shall often write $k(t, u)$ in place of $k_u(t)$, and also occasionally speak of k_u as the r.k. By virtue of F. Riesz' theorem, the r.k. of a Hilbert function space is unique.

6.2.2 Basic properties of the r.k.

a) $k(x, y) = (k_y, k_x)$

(Obvious from the definition.) Consequently k has Hermitian symmetry, i.e.

b) $k(x, y) = \overline{k(y, x)}$

c) $k(x, x) \geqslant 0$

(For, $k(x, x) = (k_x, k_x)$; note that strict inequality holds, unless $k_x = 0$, i.e. all elements of H vanish at x.)

d) $|k(x, y)| \leqslant k(x, x)^{1/2}\, k(y, y)^{1/2}$

(Follows from a), c) and Schwarz' inequality.)

e) k is positive definite (other terms used in the literature are "of positive type", "a Hermitian positive kernel", "a positive Hermitian form", etc.) on Q, i.e. for every $n \geqslant 1$ and every set of points $x_1, \ldots, x_n \in Q$, the matrix

$$\|k(x_j, x_i)\|, \qquad i, j = 1, 2, \ldots n$$

is positive semi-definite, i.e.

$$\sum_{i,j=1}^{n} k(x_j, x_i)\, \lambda_i\, \overline{\lambda_j} \geqslant 0$$

for all complex $\lambda_1, \ldots \lambda_n$.
Indeed, because of a) the determinant of $\|k(x_j, x_i)\|$ is just the Gramian of $k_{x_1}, \ldots$ $\ldots k_{x_n}$ or what is the same thing, the Hermitian form associated with this matrix equals $\|\sum_{i=1}^{n} \lambda_i k_{x_i}\|^2$ which is clearly $\geqslant 0$. Observe that we have strict positivity, unless some collection of point evaluations is linearly dependent (if this never happens, we say k is strictly positive definite).

If H is a Hilbert function space, then so is every closed subspace J thereof, which consequently also has its own r.k.. The following simple result is fundamental for the application of the r.k. in approximation theory.

6.2.3 Theorem. Let H be a Hilbert function space with reproducing kernel k_y and let H′ be a closed subspace of H with reproducing kernel k'_y. Then, the function f′ whose value at $y \in Q$ is given by

$$(1) \qquad f'(y) = (f, k'_y), \qquad f \in H$$

is an element of H′, and f′ is the orthogonal projection of f on H′. Moreover, k'_y is the orthogonal projection of k_y on H′.

Proof. Let $f \in H$ and let f_o be the orthogonal projection of f on H′, i.e. $f_o \in H'$ and $f - f_o \perp H'$. Then, since $k' \in H'$, we have for $y \in Q$

$$f'(y) = (f, k'_y) = (f - f_o, k'_y) + (f_o, k'_y) = f_o(y) ,$$

proving the first assertion. The orthogonal projection of k_y on H′ therefore takes at z the value

$$(k_y, k'_z) = \overline{(k'_z, k_y)} = \overline{k'_z(y)} = k'_y(z)$$

proving the second assertion. ◇

6.2.3.1 Corollary. If H′ and H′′ are complementary orthogonal subspaces of H, with r.k.'s k′ and k′′, then $k(x, y) = k'(x, y) + k''(x, y)$.

In applications, H is often some space $L^2(Q, \rho)$ and then

$$f'(x) = (f, k'_x) = \int f(y)\, k'(x, y)\, d\rho_y$$

gives an integral formula for the best approximation to f from H′ (examples will be given later).

An important consequence of the corollary is: if J is a subspace of K, their r.k. are related by the inequality $k^J(x, x) \leq k^K(x, x)$. It is important to remark that, in order to compute the orthogonal projection of H on H′ by formula 6.2.3 (1), it is not necessary that H possesses a r.k., but only that H′ does. (Proof unchanged.) This is important in applications where e.g. H may be $L^2(I)$ which has no r.k.!

6.2.4 Riesz' representation. In Hilbert spaces with r.k. the element representing a given bounded linear functional L can be expressed by means of the r.k..

6.2.4.1 Theorem. *Let* L *be a bounded linear functional on a Hilbert function space* H *with reproducing kernel* $k(x, y)$. *Then the representing element for* L *is*

$$h(x) = \overline{Lk_x} \quad , \tag{1}$$

i.e. the function h *defined by* (1) *is an element of* H, *and* $(f, h) = Lf$ *for all* $f \in H$.

Proof. Let h denote the (Riesz) representing element for L. Then, for all $y \in Q$,

$$Lk_y = (k_y, h) = \overline{h(y)} \ . \qquad \diamond$$

In like manner, linear operators can be represented "explicitly" by means of the r.k.. Thus, *if* A *is a bounded linear operator on* H, *then for* $f \in H$, $y \in Q$ *we have*

$$(Af)(y) = (f, A^* k_y)$$

where A^* *is the operator adjoint to* A. For

$$(Af)(y) = (Af, k_y) = (f, A^* k_y).$$

In particular, in an $L^2(Q, \rho)$ space, if the action of A^* on reproducing elements is known explicitly, then the action of A on an arbitrary element is represented by the "integral operator" $\int f(x) \, K(x, y) \, d\rho_x$ with "kernel"

$$K(x, y) = \overline{(A^* k_y)(x)} \ .$$

6.2.5 Computation of the r.k. from an orthonormal basis.

6.2.5.1 Theorem. *Let* H *be a separable Hilbert space with reproducing kernel* k *and orthonormal basis* $\{\varphi_n\}_{n=1}^{\infty}$. *Then, for all* $(x, y) \in Q \times Q$,

(1) $$k(x, y) = \sum_{i=1}^{\infty} \varphi_i(x) \overline{\varphi_i(y)} ,$$

the series converging absolutely.

Remark. Notice the remarkable fact that the sum is independent of the particular orthonormal basis chosen.

Proof. Let H_n denote the span of $\{\varphi_1, \dots \varphi_n\}$ and k^n its r.k.. Then

(2) $$k^n(x, y) = \sum_{i=1}^{n} \varphi_i(x) \overline{\varphi_i(y)} ,$$

since, writing (for fixed y) $\Phi_n = \sum_{i=1}^{n} \overline{\varphi_i(y)} \varphi_i$, $\Phi_n \in H_n$ and for every $f \in H_n$,

$$(f, \Phi_n) = \sum_{i=1}^{n} (f, \varphi_i) \varphi_i(y) = f(y)$$

because of the orthonormality of the φ_i ("Fourier decomposition" of f). Thus, by the uniqueness of reproducing elements, $\Phi_n = k^n_y$, i.e. (2) holds; therefore

$$\sum_{i=1}^{n} |\varphi_i(x)|^2 = k^n(x, x) \leq k(x, x) .$$

Letting $n \to \infty$ we see that $\sum_{i=1}^{\infty} |\varphi_i(x)|^2 < \infty$, so by Schwarz' inequality the series in (1) converges absolutely. Moreover, for fixed y the series $\sum_{i=1}^{\infty} \overline{\varphi_i(y)} \varphi_i$ converges in H, and it is immediate to check that its sum is a (and hence the unique) reproducing element for y. $\diamond$

6.2.6 The r.k. of certain special subspaces. Let $a \in Q$, and denote by H^a the set of f which vanish at a. Clearly H^a is a closed subspace of H. Let us try to find k^a, its r.k.. We assume that not all elements of H vanish at a.

The orthocomplement $(H^a)^o$ of H^a is one-dimensional. Now, the r.k. of the one-dimensional space spanned by a unit vector φ is $\varphi(x) \overline{\varphi(y)}$; hence since $(H^a)^o$ is spanned by the unit vector $k(a, a)^{-1/2} k_a$, the r.k. of $(H^a)^o$ is $k(a,a)^{-1} k_a(x) \overline{k_a(y)}$, therefore, by Corollary 6.2.3.1

$$k^a(x, y) = k(x, y) - k(a, a)^{-1}\, k(x, a)\, k(a, y)\ ,$$

which formula we can also write more suggestively

$$k^a(x, y) = \frac{\begin{vmatrix} k(x, y) & k(x, a) \\ k(a, y) & k(a, a) \end{vmatrix}}{k(a, a)}\ .$$

This formula can be generalized to give $k^n(x, y)$, the r.k. of the closed subspace $H^{a_1,\ldots,a_n}$ consisting of those functions vanishing at $a_1, \ldots a_n$. __Assuming that the point evaluations at__ $a_1, \ldots a_n$ __are linearly independent__, we have

$$k^n(x, y) = \frac{\begin{vmatrix} k(x, y) & k(x, a_1) & \ldots & k(x, a_n) \\ k(a_1, y) & k(a_1, a_1) & \ldots & k(a_1, a_n) \\ \vdots & \vdots & & \vdots \\ k(a_n, y) & k(a_n, a_1) & \ldots & k(a_n, a_n) \end{vmatrix}}{\begin{vmatrix} k(a_1, a_1) & \ldots & k(a_1, a_n) \\ \vdots & & \vdots \\ k(a_n, a_1) & \ldots & k(a_n, a_n) \end{vmatrix}}\ .$$

Indeed, the function defined by the ratio of the determinants is (for fixed y) an element of $H^{a_1,\ldots a_n}$, and since it differs from k_y by a linear combination of $k_{a_1}, \ldots k_{a_n}$ it is a (and hence __the__) reproducing element for y in $H^{a_1,\ldots a_n}$.

__Exercise__. Let $h_1, \ldots h_n$ be linearly independent elements of H. Derive a formula for the r.k. of that subspace of H consisting of all f such that $(f, h_i) = 0$, $i = 1, \ldots n$.

__6.2.7 Extremal property of the r.k.__ The following important proposition is an immediate consequence of Schwarz´ inequality.

6.2.7.1 Theorem. If H is a Hilbert space with reproducing kernel k, then for all $f \in H$ and $y \in Q$

$$|f(y)| \leq k(y, y)^{1/2} \|f\| ,$$

equality holding if and only if f is a scalar multiple of k_y.

More generally, we have

6.2.7.2 Theorem. If H is a Hilbert space with reproducing kernel k, and L is a bounded linear functional, then for all $f \in H$

$$|Lf| \leq (L(\overline{Lk_y}))^{1/2} \|f\| ,$$

equality holding if and only if f(y) is a scalar multiple of $\overline{Lk_y}$ (for notation, cf. 6.2.4.1).

Proof. Letting h denote the representing element for L, we have

$$|Lf|^2 = |(f, h)|^2 \leq (f, f)(h, h)$$

and since (h, h) = Lh, and $h(y) = \overline{Lk_y}$ by 6.2.4.1 the result follows. ◇

Example. Taking as H the Hardy space H^2 with the usual inner product, the r.k. is k_ζ: $k_\zeta(z) = k(z, \zeta) = (1 - \bar{\zeta}z)^{-1}$. Theorem 6.2.7.1 gives: for $f \in H^2$ we have $|f(\zeta)| \leq (1 - |\zeta|^2)^{-1/2} \|f\|$. Applying 6.2.7.2 with $Lf = f'(a)(|a| < 1)$ we have $\overline{Lk_\zeta} = \zeta(1-\bar{a}\zeta)^{-2}$, hence $L(\overline{Lk_\zeta}) = (1 + \bar{a}a)(1 - \bar{a}a)^{-3}$, giving the (best possible) inequality

$$|f'(a)| \leq [(1 + |a|^2)(1 - |a|^2)^{-3}]^{1/2} \|f\|$$

for all $f \in H^2$. (These results are certainly not deep, but use of the r.k. makes possible elegant computations.)

Exercise. Compute the norms of the functionals $f''(a)$, $f(a) - f(b)$ in H^2.

6.2.8 Connection of r.k. with interpolation problems. Suppose given a complex-valued function F, defined on a subset Q_o of Q, and let H_F denote the (possibly empty) set of $f \in H$ satisfying $f|_{Q_o} = F$. We have then:

6.2.8.1 Theorem. H_F is non-empty if and only if it contains an element f^* belonging to the closed span of $\{k_y, y \in Q_o\}$. In this case, f^* is the (unique) element of minimal norm in H_F.

Proof. H_F is closed and convex, and hence (if non-empty) contains a unique element g of minimal norm. Let J denote the closed span of $\{k_y, y \in Q_o\}$, J^o its orthocomplement. If h is any element of J^o, h vanishes on Q_o; hence (by the minimal property of g) $\|g + h\| \geqslant \|g\|$. This implies g is orthogonal to J^o; hence $g \in J$. ◇

Remarks. a) Note that $H_F \cap J$ contains at most one element, since if it contained two, their difference would belong to J and also vanish on Q_o, i.e. belong to J^o, implying that it is zero.

b) In the case when Q_o consists of n points $\{x_i\}_1^n$, the solution of the minimal interpolation problem leads to a system of n linear equations with the matrix $\|k(x_i, x_j)\|_{i,j=1}^n$, which is non-singular if the evaluation functionals corresponding to the x_i are linearly independent.

6.2.8.2 Universal interpolation sequences. Given a Hilbert function space H on Q with reproducing kernel k, a sequence $\{x_n\}$ of elements of Q is said to be a universal interpolation sequence (u.i.s.) if the interpolation problem

$$f(x_n) = \lambda_n , \qquad n = 1, 2, \ldots$$

is solvable whenever $\sum |\lambda_n|^2 k(x_n, x_n)^{-1} < \infty$ (for the background, and discussion of this problem, see SHAPIRO & SHIELDS$_1$). It can be shown that the necessary and sufficient condition for $\{x_n\}$ to be a u.i.s. is that the eigenvalues of the sections of the infinite Hermitian matrix $\|a_{ij}\|$ be bounded away from zero, where

$$a_{ij} = k(x_i, x_i)^{-1/2} k(x_j, x_j)^{-1/2} k(x_i, x_j) .$$

This matrix has ones on the main diagonal, and all entries bounded by one. If, in particular,

$$\sum_{j:j\neq i} |a_{ij}| \leqslant 1 - c , \qquad i = 1, 2, \ldots$$

for some $c > 0$, then the above condition is satisfied and $\{x_n\}$ is a u.i.s..

Exercise. Prove that $x_n = 1 - 2^{-n}$, $n = 1, 2, \ldots$ is a u.i.s. in the Hardy space H^2.

A situation where u.i.s. can often be constructed is that where H is a Hilbert function space defined on Euclidean n-space R^n, and translation invariant (the considerations which follow apply equally when R^n is replaced by any group, or even a set upon which a group of transformations acts, provided H has the appropriate invariance); that is, the change of variable $x \to x + \xi$ induces an isometry of H on itself. It is readily seen that then $k(x, y) = K(x - y)$ for a certain function K (which is then a positive definite function in Bochner's sense, cf. KATZNELSON p. 137). We now have:

6.2.8.3 Let G denote any countable subgroup of R^n, and suppose

$$\sum_{\substack{x \in G \\ x \neq 0}} |K(x)| < K(0)$$

Then, every square-summable sequence of prescribed values on G is taken on by a suitable element of H.

Proof. Since $k(x, x) = K(0)$ is independent of x, the assertion is just that G is a u.i.s. for H. In the above matrix, because of the group property, the sum of the absolute values of the off-diagonal elements in each row does not exceed

$$K(0)^{-1} \sum_{\substack{x \in G \\ x \neq 0}} |K(x)|$$

which, under the stated hypothesis, is less than 1. ◇

6.2.9 Examples

6.2.9.1 Let $H = \mathcal{T}_n$, i.e. the trigonometric polynomials of degree n, considered as a subspace of $L^2[-\pi, \pi]$, with inner product $(f, g) = (2\pi)^{-1} \int_{-\pi}^{\pi} f(x)\,\overline{g(x)}\,dx$. The evaluation functionals are bounded (see Exercise a) below). An orthonormal basis for this space is $\{e^{i\nu x}\}$, $\nu = -n, \ldots, n$, and the reproducing kernel is therefore

$$k(x, y) = \sum_{\nu=-n}^{n} e^{i\nu x} \cdot e^{-i\nu y} = D_n(x - y) ,$$

where

$$D_n(x) = \frac{\sin ((2n + 1) x/2)}{\sin (x/2)}$$

is the "Dirichlet kernel". (The fact that $k(x, y)$ is a "difference kernel", i.e. a function of $x - y$, is due to the underlying space $\mathcal{T}_n$ being a translation-invariant space defined on a group (here the "circle group").) The best least-square approximation to $f \in L^2[-\pi, \pi]$ from $\mathcal{T}_n$ is

$$f_n(x) = \frac{1}{2\pi} \int_{-\pi}^{\pi} f(u) D_n(x - u) du ,$$

which of course is the partial sum of rank n of the Fourier series of f.

Exercises. a) Let Q be a closed bounded interval on the real line, $\varphi_1, \ldots, \varphi_n$ linearly independent continuous functions on Q, and H the set of linear combinations of the φ_i. Prove that

(i) H is closed in $L^2(Q, dx)$

(ii) H is a Hilbert function space.

b) Considering $H = \mathcal{P}_n$ as a subspace of $L^2(I, d\rho)$, compute the r.k. of H for (i) $d\rho = dx$, (ii) $d\rho = (1 - x^2)^{-1/2} dx$ (use the Christoffel-Darboux formula, cf. DAVIS p. 238, TRICOMI$_1$ p. 127).

6.2.9.2 Let H consist of the real-valued, absolutely continuous functions f such that f and f′ belong to the Lebesgue space $L^2(R)$ and let the inner product in H be

$$(f, g) = \int_{-\infty}^{\infty} (f(x) g(x) + f'(x) g'(x)) dx .$$

(Verify that the evaluation functionals are bounded!) Clearly, the r.k. must satisfy

$$f(a) = \int_{-\infty}^{\infty} (f(x) k_a(x) + f'(x) k_a'(x)) dx = \int_{-\infty}^{\infty} f(x)(k_a(x) - k_a''(x)) dx$$

so the r.k. must be that solution of

$$k_a(x) - k_a''(x) = \delta(x - a)$$

(δ = "Dirac function") that lies in H. It is easily checked that $k_a(x) = (1/2)e^{-|x-a|}$ is the solution in question, hence $k(x,y) = (1/2)e^{-|x-y|}$. (Note that $k(x,y)$ is again a difference kernel, reflecting the translation invariance of H.) By the previous results we then know e.g. that the function in H which passes through n given points (a_i, λ_i) and has minimal norm is that linear combination of the n functions $e^{-|x-a_i|}$ which does the interpolation.

<u>Remarks</u>. For more elaborate examples of this kind, see GOLOMB & WEINBERGER, p. 146, where applications are given to problems of "optimal estimation".

In a similar manner, the r.k. in various "Sobolev spaces" of functions of several variables can be expressed in terms of the <u>fundamental solution</u> of an associated (partial) differential equation. The relation between a quadratic functional and a certain associated differential equation, which we encountered in the above example in a very simple case, is of fundamental significance for the application of variational methods to the solution of boundary value problems.

<u>Exercise</u>. Suppose $c > \log 3$ and $\{\lambda_n\}_{n=-\infty}^{\infty}$ is any square summable sequence. Prove there exists an $f \in L^2(R)$, absolutely continuous and with derivative in $L^2(R)$, such that $f(nc) = \lambda_n$, $n = 0, \pm 1, \ldots$.

<u>6.2.9.3</u> Let D be a domain in the complex plane. Then the class B(D) of functions holomorphic and square integrable in D can be shown to be a closed subspace of $L^2(D) = L^2(D, dxdy)$ endowed with the usual inner product

$$(f, g) = \iint_D f(z)\, \overline{g(z)}\, dx\, dy\ , \qquad z = x + iy\ ,$$

and the evaluation functionals in B(D) are bounded (exercise: prove these facts). Thus B(D) possesses a reproducing kernel $k_D(z, \zeta)$, called the <u>Bergman kernel</u> for the domain D. In the case when D is the unit disc $\{z: |z| < 1\}$, an orthonormal basis is given by $\varphi_n(z) = ((n + 1)/\pi)^{1/2} z^n$, $n = 0, 1, \ldots$; hence

$$k_D(z, \zeta) = \pi^{-1} \sum_{n=0}^{\infty} (n + 1) z^n \bar{\zeta}^n = \pi^{-1}(1 - \bar{\zeta}z)^{-2} .$$

Exercises. a) Show that B(D) contains non-trivial functions if the complement of D contains a disc. (Can you prove a stronger assertion?)

b) Let D be a simply connected domain in the z-plane and let $w = f(z)$ map D conformally onto the unit disc of the w-plane. Prove that

$$k_D(z, \zeta) = \frac{f'(z)\, \overline{f'(\zeta)}}{\pi(1 - f(z)\, \overline{f(\zeta)})^2} .$$

(See BERGMAN for further properties of k_D, especially its relation to Green's function, other classical domain functions, and conformal mapping.)

6.2.9.4 Let $H^2(D)$ be the Hardy space of functions on the open unit disc D with the usual inner product, i.e.

$$(f, g) = \lim_{r \to 1} (1/2\pi) \int_{-\pi}^{\pi} f(re^{i\theta})\, \overline{g(re^{i\theta})}\, d\theta .$$

It is easy to check that this is a Hilbert function space on D whose r.k. is $k(z, \zeta) = (1 - \bar{\zeta}z)^{-1}$. (The r.k. of an analogous Hilbert space formed with respect to an arbitrary domain is called the Szegö kernel of that domain; the Szegö kernel has close ties to that of Bergman and plays an important role in the complex function theory within a given domain, mapping theorems, etc.)

Denoting by P^+ the upper half-plane, $P^+ = \{z = x + iy: y > 0\}$, we have an analog of the Hardy space, which we denote by $H^2(P^+)$, consisting of those f holomorphic in P^+ such that $2\pi \|f\|^2 = \sup_{y > 0} \int_{-\infty}^{\infty} |f(x + iy)|^2\, dx$ is finite. This can be shown to be a Hilbert function space, and isometric (via the Fourier-Plancherel isometry) to $L^2(0, \infty)$. The reproducing kernel is $k(z, \zeta) = i(z - \bar{\zeta})^{-1}$. (See TITCHMARSH, Chapter V. The "reproducing" formula is here essentially the Cauchy integral formula for a half-plane.)

Exercises. a) Calculate the norms of the functionals $f^{(r)}(a)$ (where Im $a > 0$) in

$H^2(P^+)$.

b) Let S denote the horizontal strip $\{|y| < c\}$, and consider the set $H^2(S)$ of functions f such that $\|f\|^2 = \sup_{|y|<c} \int_{-\infty}^{\infty} |f(x+iy)|^2 \, dx$ is finite. Show this is a Hilbert function space, and compute the r.k..

c) Same problem for the Bergmann spaces of P^+ and S.

d) Using kernel functions, compute the orthogonal projection of $(e^{i\theta} - a)^{-1}$ (where $|a| < 1$) on H^2 (considered now as a subspace of $L^2(T)$).

e) Compute explicitly the r.k. of that subspace of H^2 consisting of all functions that vanish at n given distinct points of $\{|z| < 1\}$. (Hint: use Blaschke products).

f) In the Bergman space of the unit disc, compute the norms of the functionals f(a), f'(a) where $|a| < 1$. Also, find the maximum of $|f(b)|$ subject to the constraints $f(a) = 0$, $\|f\| = 1$.

g) In H^2, maximize $|f(b)|$ subject to the constraints $\|f\| = 1$, $f(a_1) = \ldots \ldots f(a_n) = 0$.

h) Prove that there exists a universal interpolating sequence for the Bergman space of $\{|z| < 1\}$.

i) Compute the orthogonal projection of $(z - a)^{-1}$ (where $|a| < 1$) on the Bergman space of $\{|z| < 1\}$.

6.2.9.5 Let $H = PWE_\tau$, the "Paley-Wiener" space of entire functions of exponential type $\leq \tau$ (that is, for each $\varepsilon > 0$ there is an A_ε such that $|f(z)| \leq A_\varepsilon e^{(\tau+\varepsilon)|z|}$) whose restriction to R belongs to $L^2(R)$. The inner product is

$$(f, g) = \int_{-\infty}^{\infty} f(x) \overline{g(x)} \, dx .$$

By a theorem of Paley and Wiener (cf. BOAS, p. 103), PWE_τ is identical to the space of functions of the form

$$f(z) = (1/2\pi) \int_{-\tau}^{\tau} \varphi(t) e^{-itz} \, dt , \qquad \varphi \in L^2[-\tau, \tau] .$$

It is not hard to show that PWE_τ is a Hilbert function space. By the Parseval identity we have, for each complex ζ,

$$f(\bar{\zeta}) = (1/2\pi) \int_{-\pi}^{\pi} \varphi(t)\, \overline{e^{it\zeta}}\, dt = \int_{-\infty}^{\infty} f(x) \left(\overline{\frac{\sin \tau(x-\zeta)}{\pi(x-\zeta)}} \right) dx$$

which tells us that the r.k. is

$$k(z, \zeta) = \frac{\sin \tau(z - \bar{\zeta})}{\pi(z - \bar{\zeta})}$$

An immediate consequence is the estimate, for complex $\zeta = x + iy$,

$$(1) \qquad |f(\zeta)|^2 \leqslant \|f\|^2\, k(\zeta, \zeta) = \|f\|^2 \frac{\sin 2\tau\, iy}{\pi(2iy)} = \|f\|^2 \frac{\sinh 2\tau y}{2\pi y} .$$

<u>Exercise</u>. Compute the norms of the functionals $f'(0)$ and $f''(0)$ in PWE_τ.

Let us now, for convenience, choose $\tau = \pi$. We shall deduce from (1): <u>if</u> $f \in PWE_\pi$ <u>and</u> f <u>vanishes at the integers</u>, <u>then</u> $f = 0$. Indeed, $f(z) = g(z) \sin \pi z$ for some entire g, and then (1), together with simple lower bounds for $|\sin \pi z|$, yields

$$|g(z)| \leqslant \left|\frac{f(z)}{\sin \pi z}\right| \leqslant C\, |\mathrm{Im}\, z|^{-1/2} ,$$

and (exercise!) this estimate implies $g = 0$. Hence the reproducing elements $\{k_n\}$, $n = 0, \pm 1, \ldots$ span PWE_π, and since, for $m \neq n$, $(k_m, k_n) = k_m(n) = 0$, <u>the</u> $\{k_n\}$ <u>are an orthonormal basis</u>. The "Fourier coefficients" of $f \in PWE_\pi$ relative to this basis are just $(f, k_n) = f(n)$, and we conclude: <u>for</u> $f \in PWE_\pi$,

$$(2) \qquad f(z) = \sum_{n=-\infty}^{\infty} f(n) \frac{\sin \pi(z-n)}{\pi(z-n)}$$

<u>the series converging uniformly in each strip</u> $|\mathrm{Im}\, z| \leqslant c$. The expansion (2) (called the "cardinal series") is truly remarkable, being simultaneously an expansion in orthogonal functions, and a Lagrange-type interpolation formula. It is extensively used in electrical engineering practice and other areas of applied mathematics. Observe that, <u>via</u> Fourier transformation, $\{k_n\}$ passes over to the standard orthonormal

basis $\left\{(2\pi)^{-1/2}\, e^{int}\right\}$ on $L^2(-\pi, \pi)$. The preceding analysis thus yields an exotic proof of the completeness of the trigonometric system on $L^2(-\pi, \pi)$.

6.2.9.6 As discovered by DE BRANGES, the Paley-Wiener spaces are only the simplest special cases of a general class of Hilbert spaces of entire functions. To each entire function φ which satisfies the inequality $|\varphi(x - iy)| < |\varphi(x + iy)|$ for $y > 0$ is associated the set H_φ of entire functions f satisfying

$$\|f\|_\varphi^2 = \int_{-\infty}^{\infty} |f(x)/\varphi(x)|^2 \, dx < \infty$$

together with some supplementary growth restrictions, and it can be shown that H_φ is a Hilbert function space. The reproducing kernel for H_φ takes the form

$$k_\varphi(z, \zeta) = \frac{\bar{\varphi}_2(\zeta)\ \varphi_1(z) - \bar{\varphi}_1(\zeta)\ \varphi_2(z)}{\pi(z - \bar{\zeta})}$$

where φ_1 and φ_2 are the entire functions, real-valued on the real axis, satisfying $\varphi(z) = \varphi_1(z) + i\varphi_2(z)$. For $\varphi(z) = e^{-i\tau z}$ $(\tau > 0)$, H_φ is identical with the Paley-Wiener space PWE_τ.

6.2.9.7 Consider the "Fischer space" F of entire functions f such that $\|f\|^2 = \pi^{-1} \int_0^{2\pi} \int_0^\infty |f(re^{i\theta})|^2\, e^{-r^2}\, r\, dr\, d\theta$ is finite. This is a Hilbert function space with a remarkable r.k., namely $k(z, \zeta) = \exp(\bar{\zeta} z)$. The analogous spaces in n complex variables have been studied (cf. BARGMANN) in connection with the quantum-mechanical canonical operators, and by NEWMAN & SHAPIRO$_{4,\ 5}$ in connection with convolution operators.

Exercise. Let $\{z_n\}$ be a sequence of complex numbers such that, for some $\delta > 0$

$$\sum_{j \neq i} \exp\left(-\tfrac{1}{2}\, |z_i - z_j|^2\right) \leq 1 - \delta\ , \qquad i = 1, 2, \ldots\ .$$

Then $\{z_n\}$ is a universal interpolation sequence for F. (In particular, any point lattice, reduced by a suitable scale factor, is a u.i.s.). (Hint:

$$k(\alpha, \alpha)^{-1/2}\, k(\beta, \beta)^{-1/2}\, |k(\alpha, \beta)| = \exp\left(-\tfrac{1}{2}\, |\alpha - \beta|^2\right).)$$

6.2.10 Müntz' theorem

6.2.10.1 General remarks. Even though the Hilbert space in which some particular problem is posed is not a Hilbert function space (for example, it may be $L^2(R)$, as in the example we shall consider below), there may exist an isometry onto some Hilbert function space, which relates in a natural way to the problem at hand. For instance, suppose we have a Hilbert space H_0 and a set of linearly independent elements $\{\varphi_\lambda\}$, which spans H_0, indexed by a complex parameter λ, $\lambda \in Q$. If we wish to study whether a given subset $\{\varphi_\lambda\}$, $\lambda \in Q_0$ spans H_0, it is natural to look for an isometry taking H_0 onto some Hilbert function space H on Q such that the φ_λ correspond to the reproducing elements k_λ in H. For then the original problem is transformed into that of determining whether $\{k_\lambda\}$, $\lambda \in Q_0$ span H, i.e. whether Q_0 is a set of uniqueness for H (i.e. whether an $f \in H$ which vanishes on Q_0 must vanish identically). In principle such a transformation is always possible, because the function $(\lambda, \lambda') \rightarrow (\varphi_\lambda, \varphi_{\lambda'})$ from $Q \times Q \rightarrow \not{C}$ is positive definite and every strictly positive definite function can be conceived, in a canonical way, as the r.k. of an appropriate Hilbert function space (we'll discuss this point further later). So the question is really a practical one, whether the Hilbert function space H is nice to work with, whether we can get a handle on its sets of uniqueness.

6.2.10.2 Completeness of exponentials on a half-line. Let us illustrate these remarks when H_0 is taken to be the Lebesgue space $L^2(0, \infty)$, i.e. the closed subspace of $L^2(R)$ consisting of functions vanishing for negative values. By the Fourier-Plancherel isometry, H_0 is mapped (according to a theorem of Paley and Wiener) onto (the boundary values of) the space $H^2(P^+)$ discussed earlier in 6.2.9.4. Let $\varphi_\lambda \in H_0$ be defined for each λ with Re $\lambda > 0$ by $\varphi_\lambda(t) = e^{-\lambda t}$. Since

$$\int_0^\infty \overline{e^{-\lambda t}}\, e^{itx}\, dt = (\overline{\lambda} - ix)^{-1} = i(x - \overline{i\lambda})^{-1} = k(x, i\lambda) ,$$

the Plancherel isometry thus assigns to each φ_λ, where Re $\lambda > 0$, the reproducing element $k_{i\lambda}$ in $H^2(P^+)$.

Given now a sequence $\Lambda = \{\lambda_n\}_{n=1}^{\infty}$ of points with $i\lambda_n \in P^+$ we ask: do the exponentials $\{e^{-\lambda_n t}\}$ span $H_o = L^2(0, \infty)$? By the above, the answer is: <u>if and only if $i\Lambda$ is a set of uniqueness for $H^2(P^+)$</u>.

In order to obtain a more explicit solution, let us fix an integer N, and try, for an arbitrary λ with $\mathrm{Re}\,\lambda > 0$, to compute d_N, the distance from φ_λ to the span of $\varphi_{\lambda_1}, \ldots, \varphi_{\lambda_N}$ or, what is the same thing, the distance from k_ζ to the span K_N of $k_{\zeta_1}, \ldots, k_{\zeta_N}$ where $\zeta = i\lambda$, $\zeta_n = i\lambda_n$. If k_ζ^N denotes the r.k. of K_N, we have by 6.2.3 and 6.2.3.1

$$d_N^2 = \|k_\zeta - k_\zeta^N\|^2 = \|r_\zeta^N\|^2 = r^N(\zeta, \zeta)$$

where r^N is the reproducing kernel of the orthocomplement K_N^o of K_N, i.e. of the subspace satisfying $f(\zeta_1) = \ldots f(\zeta_N) = 0$. Consider now the Blaschke product

$$B_N(z) = \prod_{n=1}^{N} \frac{z - \zeta_n}{z - \bar{\zeta}_n} .$$

Since, for fixed w, $\overline{B_N(w)}\, B_N\, k_w$ lies in K_N^o and is a reproducing element for w, it is identical with r_w^N, hence

$$r^N(\zeta, \zeta) = \overline{B_N(\zeta)}\, B_N(\zeta)\, k(\zeta, \zeta) = (2\ \mathrm{Im}\ \zeta)^{-1}\ |B_N(\zeta)|^2 ,$$

hence finally

$$d_N = (2\ \mathrm{Im}\ \zeta)^{-1/2} \prod_{n=1}^{N} \left|\frac{\zeta - \zeta_n}{\zeta - \bar{\zeta}_n}\right| .$$

Now, one can check (exercise!) that for all choices of ζ distinct from all the ζ_n, the infinite products $\prod_{n=1}^{\infty} |\zeta - \bar{\zeta}_n|^{-1}\ |\zeta - \zeta_n|$ converge or diverge together. If one of these products diverges (i.e. the partial products tend to zero), therefore every k_ζ belongs to the closed span of $\{k_{\zeta_n}\}$, and so this span is all of $H^2(P^+)$. Passing to the original problem in H_o we can state, therefore

(i) <u>The distance in</u> $L^2(0, \infty)$ <u>from</u> $e^{-\lambda t}$ to the span of $e^{-\lambda_1 t}, \ldots e^{-\lambda_N t}$ is

$$(2\ \mathrm{Re}\ \lambda)^{-1/2} \prod_{n=1}^{N} \left| \frac{\lambda - \lambda_n}{\lambda + \bar{\lambda}_n} \right|$$

(ii) <u>The necessary and sufficient condition that</u> $\{e^{-\lambda_n t}\}_{n=1}^{\infty}$ <u>span</u> $L^2(0, \infty)$ <u>is that</u>

(1) $$\lim_{N \to \infty} \prod_{n=1}^{N} \left| \frac{\lambda - \lambda_n}{\lambda + \bar{\lambda}_n} \right| = 0$$

<u>for some (and hence every) choice of</u> λ <u>distinct from all the</u> λ_n.

These assertions contain the essence of a celebrated theorem of Müntz. It is a remarkable phenomenon that the closed span of a given sequence of exponentials either picks up <u>no</u> new exponentials at all, or else it picks up <u>all</u> of them (and the whole space $L^2(0, \infty)$).

If we take λ and the λ_n to be real and satisfy $0 < \lambda$, $0 < \lambda_1 < \lambda_2 < \ldots$, then (1) holds if and only if $\sum \lambda_n^{-1} = \infty$. Moreover, by means of the change of variable $x = e^{-t}$ we can transform the result into one concerning the span of $\{x^{\lambda_n}\}$ in $L^2(0, 1)$. (The details are left as an exercise.) We thus obtain an L^2 version of Müntz´ theorem:

<u>Necessary and sufficient condition that the linear span of the functions</u> $x^{\lambda_1}, x^{\lambda_2}, \ldots$, where $-(1/2) < \lambda_1 < \lambda_2 < \ldots$, <u>is dense in</u> $L^2(0, 1)$ <u>is</u> $\sum (1+\lambda_n)^{-1} = \infty$.

To get the standard ("uniform") version one uses the inequality

$$\max_{0 \leq x \leq 1} |f(x)| \leq f(0) + \left(\int_0^1 |f'(x)|^2 \, dx \right)^{1/2}$$

and obtains (the details are left to the reader):

<u>Necessary and sufficient condition that the linear span of the functions</u> $1, x^{\lambda_1}, x^{\lambda_2}, \ldots$, <u>where</u> $0 < \lambda_1 < \lambda_2 < \ldots$, <u>is dense in</u> $C[0, 1]$ <u>is</u> $\sum \lambda_n^{-1} = \infty$.

<u>6.2.10.3 Remarks</u>. The proof we have outlined above (the backbone of which is the formula for d_N) is not the most elementary, based as it is upon Plancherel´s theorem and the Paley-Wiener theorem for the half-line; we have given this proof here because of the instructive computations with kernel functions. The most direct proof (Müntz´ method) is to start from the formula for d_N^2 as a quotient of two Gram determinants, which in the present case are easy to evaluate.

A variant of the above proof can also be given which avoids the Paley-Wiener theorem and $H^2(P^+)$, and by a rather more recondite isometry (based upon Laguerre functions) throws the problem back on the ordinary Hardy space H^2 of the unit disc (SHAPIRO & SHIELDS$_2$ p. 227).

One final methodological remark: The use of Fourier transforms in the above proof can also be avoided. Let us first make the general observation that <u>in any Hilbert space, the distance from</u> x_0 <u>to the space spanned by</u> $x_1, \ldots, x_N$ <u>is a function only of the inner products</u> (x_i, x_j), $i, j = 0, 1, \ldots, n$. Now, the inner product in $L^2(0, \infty)$ of $e^{-\lambda_i t}$ and $e^{-\lambda_j t}$ is

$$\int_0^\infty e^{-\lambda_i t} e^{-\overline{\lambda}_j t} dt = (\lambda_i + \overline{\lambda}_j)^{-1} = (k_{\lambda_j}, k_{\lambda_i})$$

where $k(z, \zeta) = (z + \overline{\zeta})^{-1}$ is the r.k. of the H^2 space of the right half-plane (i.e. the -90° rotation of P^+). Therefore, the computation of d_N is immediately reduced to the analogous computation for the reproducing elements, as done above. It is a remarkable fact that, modulo elementary computations, the Müntz theorem is thus implicit in the basic properties of the space $H^2(P^+)$.

<u>6.2.11 Concluding remarks about r.k.</u>

<u>6.2.11.1 Relation to positive definite functions.</u> We have seen that if $k(s, t)$ is the r.k. of a Hilbert function space H of functions defined on a set Q, then $k(s, t)$ is positive definite on $Q \times Q$. It is an important fact that, conversely, every (strictly) positive definite function k on $Q \times Q$ determines a unique Hilbert function space on Q, so that the concepts "Hilbert function space" and "positive definite function" are essentially equivalent (cf. ARONSZAJN). The idea, which goes back to early work of E.H. Moore, is as follows.

Given the strictly p.d. function k on $Q \times Q$, consider, for each $t \in Q$, the function $k_t: s \rightarrow k(s, t)$ from Q to $\mathbb{C}$, and define H_0 to be the linear manifold consisting of all finite linear combinations of the elements $\{k_t\}_{t \in Q}$. We introduce an inner product (,) on H_0 by defining $(k_t, k_{t'}) = k(t', t)$ for $t, t' \in Q$ and extending to all of H_0 by sesquilinearity. The strict positive definiteness of k ensures that this bilinear form has the crucial properties of Hermitian symmetry, and positivity (i.e. $(f, f) > 0$ if $f \neq 0$), so we have indeed an inner product on H_0. We now complete H_0 with respect to the norm induced by this inner product, obtaining a Hilbert space H. Because of the readily verified relation $|f(t)| = |(f, k_t)| \leq k(t,t)^{1/2} \|f\|$ for $f \in H_0$, a sequence of elements of H_0 which converges in this norm converges pointwise on Q; hence the elements of H can be interpreted as complex-valued functions on Q, the point evaluations being bounded. It is easy to check that H is then a Hilbert function space on Q whose r.k. is $k(s, t)$.

This procedure of constructing a Hilbert space out of a positive definite function has been widely used. It is commonly employed in the spectral theory of operators (for references, see e.g. the notes to Chapter 8 of BEREZANSKI); it has been employed by Sz.-Nagy and Korànyi to obtain an elegant solution of the Pick-Nevanlinna interpolation problem (references in AHIEZER_2).

6.2.11.2 The reader interested in learning more about Hilbert spaces with r.k. may consult (among many others) the following sources: For general background, ARONSZAJN (also relevant material in DAVIS and CHALMERS); for applications to numerical analysis and approximation, DAVIS; cf. also GOLOMB & WEINBERGER; for the Bergman and Szegö kernels, BERGMAN and MESCHKOWSKI; for PWE_T, BOAS and DE BRANGES, where far-reaching generalizations are given; for interpolation problems, SHAPIRO & $\text{SHIELDS}_{1,2}$, ROSENBAUM and DUREN & WILLIAMS; SHAPIRO_5 gives a proof using r.k. of (a generalization of) Beurling's invariant subspace theorem, as well as applications to extremal and uniqueness problems for bounded analytic functions. In closing this section we wish to express the view that the study of Hilbert spaces with r.k. (and perhaps, generalizations to Banach function spaces with bounded point evaluations) still offers an interesting field for exploration. Perhaps there is not too much general theory yet to be developed, but there are concrete spaces to study, and what is more important, one could seek novel applications of Hilbert spaces with r.k.. Because of various isometries, spaces with r.k. may well be useful even in the study of problems in Hilbert spaces without r.k. (we saw this in the case of Müntz' theorem). Another case in point is the important role played by the Hardy space H^2 in least-squares prediction theory (GRENANDER & SZEGÖ, LEVINSON & MCKEAN, also MASANI and references there), and just as this chapter was completed we observed that DYM & MCKEAN have made similar application of de Branges' spaces. As a final instance, the Fischer space F can be a useful vehicle for the study of Fourier transforms on $L^2(R)$; for instance, under an isometry from $L^2(R)$ onto F given by BARGMANN, the Fourier-Plancherel transformation on $L^2(R)$ is unitarily equivalent to the rather harmless transformation in F induced by rotation of the independent variable through a right angle (this observation is essentially equivalent to Wiener's proof (cf. TITCHMARSH) of the Plancherel theorem by means of Hermite functions).

6.3 Quadratic Functionals. An important type of problem is that of maximizing a quadratic functional on the unit sphere of a Hilbert space. (In its finite-dimensional version, this is just the problem of the maximum of a Hermitian form, which is the largest eigenvalue of its associated matrix.) Suppose, on an abstract Hilbert space H, we are given a Hermitian form, that is, a map $(\ , \ )_0$ from $H \times H$ to $\mathbb{C}$ which is linear in the first variable and satisfies $(f, g)_0 = \overline{(g, f)_0}$, i.e. has Hermitian symmetry. Then the functional $J(x) = \|x\|_0^2 = (x, x)_0$ is called a quadratic functional on H. Here we assume that $J(x)$ is bounded on the unit sphere of H. A situation of common occurrence is that where the function $x \to J(x)$ is continuous when H is endowed with its weak topology. This is sufficient to ensure that $J(x)$ attains a maximum, say $J(x^*) = M$, for at least one point on the (compact, for this topology) unit sphere of H. (An example is gotten by taking H to be the "Bergman space" of functions f holomorphic in the open unit disc D with $\|f\|^2 = \int_D |f(z)|^2 \, d\sigma < \infty$ (σ: planar Lebesgue measure), and $J(f) = \int_\Delta |f(z)|^2 \, d\sigma$ where Δ is a domain whose closure lies in D.) The "variational conditions" characterizing a maximizing element for a quadratic functional are given by

6.3.1 Theorem. *Let $(\ , \ )_0$ be a Hermitian form on the Hilbert space H, whose inner product is written $(\ , \ )$, and J the quadratic functional induced by $(\ , \ )_0$. If $J(x)$ attains a maximum value M, say $J(x^*) = M$, as x ranges over the unit sphere of H, then*

$$(1) \qquad (x^*, x)_0 = M(x^*, x) , \qquad \text{all } x \in H .$$

Proof. Clearly we may suppose $M > 0$. Let y be any element of H such that

$$(x^*, y)_0 = 0 \tag{2}$$

Then, since $J(x^* + y) \geqslant J(x^*)$, we must have $\|x^* + y\| \geqslant \|x^*\|$, or else, because of homogeneity, we should contradict the maximality of x^*. Replacing y by λy, with λ a complex scalar (λy also satisfies (2)) shows that $\|x^* + \lambda y\| \geqslant \|x^*\|$, and hence, since λ is arbitrary, $(x^*, y) = 0$. Now, for any $x \in H$,

$$y = (x, x^*)_0 \, x^* - Mx$$

satisfies (2), which implies (1). ◇

The condition (1) is only <u>necessary</u> for a maximizing element, i.e. if we can find x^* of norm 1 and $c > 0$ such that $(x^*, x)_0 = c(x^*, x)$ holds for all $x \in H$, it need not be true that $c = \sup_{\|x\|=1} J(x)$. To see this, let y_1 denote a maximizing element, i.e. $\|y_1\| = 1$ and $J(y_1) = M_1 = \sup_{\|x\|=1} J(x)$, and consider now the extremal problem of maximizing $J(x)$ subject to the conditions

$$\|x\| = 1 , \qquad (y_1, x)_0 = 0 .$$

Suppose y_2 is a maximizing element for this problem, and $J(y_2) = M_2$, where of course $M_2 \leqslant M_1$. The same variational argument used above shows that, for $y \in H$

$$(y_1, y)_0 = (y_2, y)_0 = 0 \quad \text{implies} \quad (y_2, y) = 0 .$$

Hence, by simple linear algebra

$$(y_2, x)_0 = M_2(y_2, x) , \qquad \underline{\text{all}} \quad x \in H .$$

Thus, y_2 also satisfies the variational condition. Proceeding in this manner, we get a sequence of elements $\{y_n\}$, $n = 1, 2, \ldots$ of norm one, and positive numbers $M_1 \geqslant M_2 \geqslant \ldots$ such that

$$(y_n, x)_0 = M_n(y_n, x) , \qquad \text{all} \quad x \in H .$$

The reader will easily recognize that we have here to do with an <u>eigenvalue problem</u>. (A concise orientation in eigenvalue theory of quadratic forms is DIAZ; see also COURANT & HILBERT, GOULD, STENGER, HAMBURGER & GRIMSHAW - although only finite-dimensional problems are dealt with in the last-named work, much of the discussion of extremal properties etc. carries over with little change to the infinite-dimensional context.)

Now we can see why (1) is not sufficient for a maximizing element: if $J(x)$ is, for example, $\|Ax\|^2$, where A is a compact operator, the numbers M_n above, which are the eigenvalues of A^*A, tend to zero as $n \to \infty$, hence $M_n < M_1$ must hold for some n.

6.3.2 Case when H is a Hilbert function space. The variational condition assumes the concrete form of an integral equation if we assume that H is a Hilbert function space, whose elements are functions defined on some measurable space Q, and

$$(f, g)_o = \int f \bar{g} \, d\rho_o$$

for some positive measure ρ_o on the Borel sets of Q.

If $k(s, t)$ denotes the r.k. of H we get, writing f_n for y_n in (4), and taking k_t for x,

$$\int f_n(s) \, k(t, s) \, d\rho_o(s) = M_n \, f_n(t) \ .$$

In other words, the M_n are eigenvalues, and the f_n eigenfunctions of the integral operator

$$A: f \to \int f(s) \, k(t, s) \, d\rho_o(s) \ .$$

This A is a Hermitian operator, and one checks easily that

$$J(f) = (f, f)_o = \int |f|^2 \, d\rho_o = (Af, f) \ .$$

This leads to another interpretation (and proof) of the results obtained above, namely: a quadratic functional on a Hilbert function space H, which has the form $\int |f|^2 \, d\rho_o$, can be represented in the form (Af, f) for a suitably defined Hermitian (integral) operator A; in particular, its supremum on the unit sphere of H is the largest eigenvalue of A. (Actually, a quadratic functional is representable in the form (Af, f) under much more general conditions, but we shall not go into this here; cf. RIESZ & NAGY.)

6.3.2.1 Examples. a) We may take for H the Paley-Wiener space PWE_T of entire functions (see 6.2.9.5), and for fixed $a > 0$, $J(f) = J_a(f) - \int_{-a}^{a} |f(x)|^2\,dx$. The problem of maximizing $J(f)$ subject to $\|f\| = 1$ arises in communication theory (the "concentration problem"). By the above, the maximum in question is the largest eigenvalue λ_{max} of the integral equation

$$\int_{-a}^{a} f(\xi)\ \frac{\sin \tau(x - \xi)}{\pi(x - \xi)}\ d\xi = \lambda f(x)$$

and the f belonging to λ_{max} are extremal functions. An "explicit solution" can be given in terms of so-called <u>spheroidal wave functions</u> (cf. LANDAU & POLLAK, SLEPIAN & POLLAK; also LOGAN for the L^p analogue).

b) Let H denote the "Bergman space" of the unit disc D, and $J(f) = \int_\Delta |f(z)|^2\,d\sigma$, where Δ is a domain whose closure lies in D. The maximum of $J(f)$ subject to $\|f\| = 1$ is the largest eigenvalue λ_{max} of

$$(1/\pi) \int_\Delta f(\zeta)(1 - \bar{\zeta}z)^{-2}\,d\sigma = \lambda f(z)$$

and the f belonging to λ_{max} are extremal functions, i.e. maximize $J(f)$.

Observe that eigenfunctions f_m, f_n belonging to distinct eigenvalues λ_m, λ_n are orthogonal in H, and hence, by virtue of the variational conditions we have derived, they are orthogonal also with respect to the inner product $(\ ,\ )_0$. This remarkable <u>double orthogonality</u> of the eigenfunctions was remarked by BERGMAN in the context of 6.2.9.3. It is really a general Hilbert space phenomenon, a consequence of the Hermitian character of the above operator A. In many typical cases, A turns out to be compact ("completely continuous") in which case further facts can be deduced, e.g. $\lambda_n \to 0$.

For discussion of eigenvalue problems for integral operators, see $TRICOMI_2$.

<u>Exercises</u>. a) Prove, for each of the above examples, that the operator A is compact; that the "doubly orthogonal" system is complete in H. In example b), show that the eigenfunctions are continuable analytically across $\{|z| = 1\}$; solve the problem explicitly when $\Delta = \{|z| \leq a\}$, $a < 1$.

b) Taking for H the Hardy space H^2 of the disc, set up the eigenvalue problem associated with the quadratic functional $J(f) = \int_0^a |f(x)|^2\, dx$, $a < 1$.

c) Again, H = Hardy space H^2 of the unit disc D. Let $z_i \in D$, $i = 1, \ldots n$, and define

$$M = \max \sum_{i=1}^{n} c_i \, |f(z_i)|^2 , \qquad f \in H^2(D) \ , \ \ \|f\| = 1 ,$$

where c_i are given positive numbers. Prove that every extremal function g satisfies

$$Mg(z) = \sum_{i=1}^{n} c_i \, g(z_i)(1 - \bar{z}_i z)^{-1}$$

d) Examine the question of uniqueness (apart from constant factors) of the extremals in the above problems.

e) In connection with example a) above, is the operator A compact if $[-a, a]$ is replaced by $[0, \infty)$?

In example b), can you find a domain Δ whose closure contains the point $z = 1$ such that the associated operator is compact? Similar question *à propos* exercise b), for the interval $[0, 1]$. In exercise c), can you find an infinite sequence $\{z_i\}$ such that, with $c_i = 1 - |z_i|^2$, the associated operator is compact?

6.3.3. Remarks. The subject of quadratic functionals is very important in mathematical physics. Here we have only dealt with one aspect of the problem (and that superficially) because of its connection with other material dealt with in this chapter. The practical consequences of the above discussion are mainly that the (physically important) eigenvalue problem for certain integral operators admits an interpretation in terms of maximizing a quadratic functional, and this problem can be attacked computationally by so-called direct methods of the calculus of variations.

One of the most basic extremal problems in mathematical physics is that of minimizing a functional of the form

$$F(x) = (Ax, x) - 2 \operatorname{Re}(x, \xi) ,$$

where ξ is a fixed element of some Hilbert space H, as x ranges over the entire space H. Here A is some (usually "positive") linear operator. In practice, H is some space of functions ("Sobolev space" or a variant thereof) on a domain in R^n, and A is a differential operator. The above extremal problem is set up so that a minimizing element x^* yields a solution of the partial differential equation $Ax = \xi$ with vanishing boundary values. This line of research, which originated in attempts to validate the "Dirichlet principle", is one of the most central, and also most beautiful areas of mathematical analysis. It is in this connection that Hilbert space theory (originally, in the guise of a general theory of integral equations) originated, and it is here too that Hilbert space theory has recorded some of its most spectacular achievements. A good introduction to these matters is MIKHLIN, cf. also COURANT.

<u>6.4 Completeness of a set of elements in Hilbert space</u>. Although we shall devote but little attention in these lectures to problems of completeness, bases, etc. (cf. the treatise of SINGER_2), we assemble here a few remarks of an elementary nature.

<u>6.4.1 Parseval's identity</u>. Let $\{y_n\}_1^\infty$ be an orthonormal system (ONS) in the Hilbert space H. Suppose we can verify that the Parseval identity

(1) $$\|x\|^2 = \sum_{n=1}^{\infty} |(x, y_n)|^2$$

holds, for a particular $x \in H$. Then, since the left side minus the right side is an expression for the square of the distance from x to the closed subspace Y spanned by the y_n, we know $x \in Y$. Hence, <u>if</u> (1) <u>holds for all</u> x <u>in some set</u> S <u>whose linear combinations are dense in</u> H, <u>then</u> $\{y_n\}_{n=1}^\infty$ <u>span</u> H. This criterion is often useful in practice. As an application we have,

<u>6.4.1.1 Theorem</u>. <u>Necessary and sufficient that the</u> ONS $\{\varphi_n\}$, $\varphi_n \in L^2(a, b)$ <u>span</u> $L^2(a, b)$ <u>is that either of the following conditions hold</u>:

(i) $$\sum_n \left|\int_a^x \varphi_n(t)\, dt\right|^2 = x - a, \qquad a \leq x \leq b$$

(ii) $$\sum_n \int_a^b |\int_a^x \varphi_n(t)\, dt|^2\, dx = (b-a)^2/2$$

Proof. (i) simply says that the Parseval identity holds for the characteristic function of $[a, x]$, and the totality of these functions spans $L^2(a, b)$. Thus, (i) is a N.A.S.C.. Moreover, the same argument clearly shows that it is sufficient if (i) holds for a set of x dense in $[a, b]$.

Now, (i) implies (ii). Finally, we show that if (ii) holds then also (i) holds for a dense set of $x \in [a, b]$. Indeed, by Bessel's inequality (applied to the characteristic function of $[a, x]$)

(1) $$(x-a) - \sum_n |\int_a^x \varphi_n(t)\, dt|^2 \geqslant 0$$

for each $x \in [a, b]$. Suppose now that (ii) holds; then the integral of the expression on the left side of (1) vanishes; hence this expression itself vanishes for a set of x that is dense in $[a, b]$; that is, (i) holds for such x. ◇

For historical remarks concerning this theorem, see TRICOMI$_1$ p. 26 ff. It is amusing to see what the above criteria say in simple situations. Let us take $\varphi_n(t) = (2\pi)^{-1/2} e^{int}$, $n = 0, \pm 1, \ldots$ and test for completeness in $L^2(0, 2\pi)$. The test (i) yields the condition

$$\sum_{n=-\infty}^{\infty} |\int_0^x e^{int}\, dt|^2 = 2\pi x, \qquad 0 \leqslant x \leqslant 2\pi$$

which we can write

(2) $$x^2 + 8\sum_{n=1}^{\infty} (\sin^2 nx/2)/n^2 = 2\pi x$$

and (ii), obtained by termwise integration, becomes (with a little manipulation)

(3) $$\sum_1^{\infty} n^{-2} = \pi^2/6 .$$

Thus, any independent verification of (3) (e.g. by residue calculus) yields an exotic proof of the density of the trigonometric polynomials in $L^2(0, 2\pi)$.

As another illustration, let us seek to prove that (ordinary) polynomials are dense in $L^2(-1, 1)$. Unfortunately, the method applies only to orthonormal systems, so we work with normalized (i.e. of norm 1) Legendre polynomials (cf. NATANSON, p. 275)

$$\varphi_n(x) = ((2n+1)/2)^{1/2}\,(1/2^n\,n!)(d^n/dx^n)(x^2-1)^n = ((2n+1)/2)^{1/2}\,P_n(x)$$

The completeness criterion (ii) becomes here

$$(4) \qquad \sum_{n=0}^{\infty} (n+\tfrac{1}{2}) \int_{-1}^{1} \Big(\int_{-1}^{x} P_n(t)\,dt \Big)^2 dx = 2$$

and using the relations $P_n(-1) = (-1)^n$, and

$$(2n+1)\,P_n(x) = (d/dx)(P_{n+1}(x) - P_{n-1}(x)), \qquad n \geqslant 1$$

(loc. cit., p. 277 - 278), (4) is easily verified. In like manner, the completeness of other systems of classical orthogonal polynomials (with respect to various weight functions) can be deduced by verifying (ii), which involves only simple computations once the simplest algebraic properties of the polynomials in question are established. (However, there are other completeness proofs which are less computational, and perhaps more enlightening.)

Exercises. a) Fill in the missing details in the above proof of completeness of the Legendre polynomials.

b) Prove that the Chebyshev polynomials span $L^2([-1, 1]; (1-x^2)^{-1/2}dx)$.

c) Deduce from the completeness of the Legendre polynomials the Weierstrass theorem that polynomials are dense in $C(-1, 1)$. (Hint: assume first f is a piecewise linear function and approximate its derivative in L^2 norm.)

d) Verify that the functions $\left\{\sqrt{2}\cos(n+\tfrac{1}{2})\pi x\right\}_{n=0}^{\infty}$ are orthonormal in $L^2(0, 1)$, and use 6.4.1.1 (ii) to prove that they are complete.

6.4.2 Perturbation of orthonormal bases. An orthonormal basis in a separable Hilbert space H is known to be "stable", in the sense that if its elements are perturbed "slightly" the resulting system, while of course not orthogonal, is still complete, and in fact admits a biorthogonal system (details below). This enables us to expand an arbitrary element of H in terms of the perturbed basis. The first results of this kind are apparently due to Paley and Wiener. Here we follow the elegant treatment of SZ. NAGY$_2$ (reproduced in RIESZ & NAGY; for further developments, and extensions to Banach spaces, see SINGER$_2$).

6.4.2.1 Definition. A sequence $\{z_n\}_{n=1}^{\infty}$ of elements of H is a Bessel system if there exists a number M such that

$$(1) \qquad \sum_{n=1}^{\infty} |(x, z_n)|^2 \leq M \|x\|^2 , \qquad \text{all } x \in H .$$

In that case, the infimum M^* of numbers M for which (1) holds is called the (Bessel) constant of the system (it is easy to verify that (1) then holds with $M = M^*$).

The reason for the designation is, when $M = 1$, (1) is just the Bessel inequality. Thus, every ONS is a Bessel system with constant 1. Simple criteria for a set $\{z_n\}$ to be a Bessel system (which will be needed later) are stated below in exercises c, d.

Exercises. a) Prove that (1) holds for $M = M^*$.

b) Prove that (1) is equivalent to the proposition: the matrix $\|(z_i, z_j)\|$ $(i, j = 1, 2, \ldots)$ defines a bounded Hermitian operator on the Hilbert space ℓ^2 of square-summable sequences with bound not exceeding M.

c) Prove that (1) is equivalent to the statement: $\|\sum_{n=1}^{N} c_n z_n\|^2 \leq M \sum_{n=1}^{N} |c_n|^2$ for all finite sequences of complex numbers $c_1, \ldots c_N$ (N arbitrary).

d) Prove that if $\sum \|z_n\|^2 < 1$, $\{z_n\}$ is a Bessel system with constant less than 1.

6.4.2.2 Theorem (Paley, Wiener, Sz. Nagy). Let $\{e_n\}_{n=1}^{\infty}$ be a complete orthonormal system in the Hilbert space H, and $\{z_n\}$ a Bessel system with Bessel constant $c^2 < 1$. Let $x_n = e_n + z_n$. Then

(i) $\{x_n\}$ is complete (i.e. spans H).

(ii) There exists a complete sequence $\{y_n\}_{n=1}^{\infty}$ biorthogonal to $\{x_n\}$ (i.e. $(x_m, y_n) = \delta_{mn}$).

(iii) Every $x \in H$ can be expressed as the sum of the norm-convergent series

(1) $$x = \sum_1^{\infty} (x, y_n)x_n = \sum_1^{\infty} (x, x_n)y_n \ .$$

(iv) For every $x \in H$ we have

(2) $$(1 + c)^{-2} \|x\|^2 \leqslant \sum_{n=1}^{\infty} |(x, y_n)|^2 \leqslant (1 - c)^{-2} \|x\|^2$$

(3) $$(1 - c)^2 \|x\|^2 \leqslant \sum_{n=1}^{\infty} |(x, x_n)|^2 \leqslant (1 + c)^2 \|x\|^2 .$$

Proof. Let us define a linear transformation A by $Ae_n = z_n$, $n = 1, 2, \ldots$. Initially it is defined only on the (dense) subset of H consisting of elements $x = \sum_1^N c_n e_n$. However, for such x, $\|Ax\|^2 = \| \sum_1^N c_n z_n\|^2 \leqslant c^2 \sum_1^N |c_n|^2 = c^2 \|x\|^2$, so that A is extendible to all of H as a bounded linear operator with norm not exceeding c. Recall now that $c < 1$. Hence $B = I + A$ is invertible, where I denotes the identity operator, and

(4) $$\|B^{-1}\| \leqslant (1 - c)^{-1} \ .$$

Now, $Ae_n = z_n$; hence $Be_n = x_n$. Define

$$y_n = (B^{-1})^* e_n \ .$$

We claim that $\{y_n\}$ has all the stated properties. Observe first that $(x_m, y_n) = (x_m, (B^{-1})^* e_n) = (B^{-1} x_m, e_n) = (e_m, e_n) = \delta_{mn}$. Now, $B^{-1} x = \sum (B^{-1} x, e_n) e_n$ and applying the operator B termwise (legitimate because the series is norm-convergent) gives $x = \sum (B^{-1}x, e_n)x_n = \sum (x, y_n)x_n$, the latter series being again norm-convergent. This establishes the first equality in (1), and as a corollary the completeness of $\{x_n\}$. Similarly, if we start from $B^*x = \sum (B^*x, e_n)e_n$ and apply the bounded operator $(B^*)^{-1} = (B^{-1})^*$ termwise we get

$$x = \sum (B^*x, e_n)y_n = \sum (x, x_n)y_n$$

establishing the second equality in (1), and consequently also the completeness of $\{y_n\}$.

All that now remains to be proved is (iv). Now, by Parseval's identity $\|B^{-1}x\|^2 = \sum |(B^{-1}x, e_n)|^2 = \sum |(x, y_n)|^2$. This, together with (4), proves the right hand inequality in (2), and the analogous reasoning based on $\|B\| \leqslant 1 + c$ (hence $\|B^{-1}\| \geqslant (1 + c)^{-1}$) establishes the left-hand inequality. Finally, $\|B^*x\|^2 = \sum |(B^*x, e_n)|^2 = \sum |(x, x_n)|^2$, and since $(1 - c)\|x\| \leqslant \|B^*x\| \leqslant (1 + c)\|x\|$ (because $B^* = I + A^*$ and $\|A^*\| \leqslant c$) we get (3). ◇

As an application, let us prove the following theorem, a refinement by DUFFIN & EACHUS of an earlier result of Paley and Wiener.

6.4.2.3 Theorem. *The functions* $\{e^{i\lambda_n t}\}$ *are complete (total) in* $L^2(-\pi, \pi)$, *and admit a biorthogonal system if* $|\lambda_n - n| \leqslant \alpha$, $n = 0, \pm 1, \dots$ *where* α *is any number less than* $(\log 2)/\pi$.

Proof. We wish to apply the preceding theorem to the complete ONS $e_n \equiv \varphi_n(t) = (2\pi)^{-1/2}\, e^{int}$ (where the index n runs now through *all* integers) and $x_n \equiv (2\pi)^{-1/2} e^{i\lambda_n t}$, and therefore must show that $z_n \equiv (2\pi)^{-1/2}\,(e^{i\lambda_n t} - e^{int})$ is a Bessel system with constant less than one. Now, if c_n is any complex sequence that is zero for $|n|$ large enough, and a_n denotes $\lambda_n - n$,

$$\sum_{n=-\infty}^{\infty} c_n z_n \equiv \sum_{n=-\infty}^{\infty} c_n \varphi_n(t)(e^{ia_n t} - 1) = \sum_{n=-\infty}^{\infty} c_n \varphi_n(t)\Big(\sum_{k=1}^{\infty} (ia_n t)^k/k!\Big) =$$

$$= \sum_{k=1}^{\infty} \left(\sum_{n=-\infty}^{\infty} a_n^k c_n \varphi_n(t) \right) (it)^k / k! = \sum_{k=1}^{\infty} (it)^k g_k(t)/k!$$

where

$$g_k(t) = \sum_{n=-\infty}^{\infty} a_n^k c_n \varphi_n(t)$$

satisfies

$$\|g_k\|^2 = \sum_{n=-\infty}^{\infty} |a_n|^{2k} |c_n|^2 \leqslant \alpha^{2k} \sum |c_n|^2 .$$

Since $t^k g_k(t)$ has norm $\leqslant \pi^k \|g_k\|$, we have

$$\left\| \sum c_n z_n \right\| \leqslant \sum_{k=1}^{\infty} (k!)^{-1} \|t^k g_k(t)\| \leqslant \left(\sum_{k=1}^{\infty} (k!)^{-1} \pi^k \alpha^k \right) \left(\sum |c_n|^2 \right)^{1/2} =$$

$$= (e^{\alpha\pi} - 1) \left(\sum |c_n|^2 \right)^{1/2}$$

and if $e^{\alpha\pi} - 1 < 1$, i.e. $\alpha < (\log 2)/\pi$, this is one of the equivalent criteria for $\{z_n\}$ to be a Bessel system with constant < 1. The assertions of the theorem (and a good deal more, which we do not state explicitly) now follow from 6.4.2.2. ◇

Remark. The best constant here is not $(\log 2)/\pi$, but $1/4$, as established by Levinson (see LEVINSON, Chapter 4, where also analogous L^p results are established; cf. also HAMMERSLEY).

6.4.2.4 We can draw further interesting consequences in connection with the Paley-Wiener Hilbert space PWE_π of entire functions, introduced in 6.2.9.5. Let us here write simply (PW) to denote this space. Now, the function $\psi_\lambda(t)$ defined as $e^{i\lambda t}$ on $[-\pi, \pi]$ and zero outside is mapped under Fourier transformation into

$$\sin \pi(\lambda - x)/\pi(\lambda - x) = k(\lambda, x) = \overline{k_\lambda(x)}$$

where k is the r.k. of (PW). Thus (cf. 6.2.10.1), problems of totality of families of (complex) exponential functions in $L^2(-\pi, \pi)$ are equivalent to problems of sets of uniqueness (equivalently, the study of zero-sets) for functions in (PW). In particular, the result just proved has the following consequences. If $\{\lambda_n\}$ satisfies

$|\lambda_n - n| \leq \alpha < (\log 2)/\pi$, $n = 0, \pm 1, \ldots$ then $\{\lambda_n\}$ is a set of uniqueness for (PW) (i.e. $f \in (PW)$, $f(\lambda_n) = 0$, $n = 0, \pm 1, \ldots$ implies f is zero). This result is rather delicate, for $\{\lambda_n\}$ is a minimal set of uniqueness, i.e. if any element is deleted from the set, the resulting set is not a set of uniqueness, or what is the same thing, the family of exponentials $\{\psi_{\lambda_n}\}$ no longer has the property of totality. The latter statement follows immediately from the existence of a biorthogonal sequence. More specifically, let $\{\varphi_n\}$ denote the (unique) biorthogonal system to $\{\psi_{\lambda_n}\}$. The inner product in $L^2(-\pi, \pi)$ will be understood as

$$(\varphi, \psi) = (2\pi)^{-1} \int \varphi(t)\, \overline{\psi(t)}\, dt$$

so that $(\psi_\lambda, \psi_\zeta) = k(\lambda, \zeta)$. If now $f_n \in (PW)$ denotes the Fourier transform of φ_n, we have (λ_n real)

$$(\varphi_m, \psi_{\lambda_n}) = (f_m, k_{\lambda_n}) = f_m(\lambda_n) = \delta_{mn}$$

(Observe that the first inner product is understood in the space $L^2(-\pi, \pi)$, the second in (PW).).

The point of all this is that via the Fourier isometry we propose to transplant 6.4.2.3 into the space (PW) and come out with a theorem about entire functions, and now finally the stage has been set. We have gotten hold of a system $\{f_n\}$ in (PW) biorthogonal to the system of reproducing elements $\{k_{\lambda_n}\}$, and therefore every $f \in (PW)$ has the norm-convergent expansions

$$f = \sum (f, f_n)\, k_{\lambda_n} = \sum (f, k_{\lambda_n})\, f_n$$

of which the latter is the more interesting, and yields

$$f(z) = \sum f(\lambda_n)\, f_n(z) \,, \tag{1}$$

the convergence being uniform on each horizontal strip $|\mathrm{Im}\, z| \leq a$. Now, (1) is a simple example of an extension into the transcendental of the Lagrange interpolation formula; formulas of this kind are common in the interpolatory theory of entire functions (they occur e.g. throughout the book of BOAS, mostly in lemmas needed for proving uniqueness and growth theorems). Moreover, since we know the zeroes of f_n it is not hard to get an explicit formula for it. We don't want to go deeper into this matter here, referring the reader rather to LEVINSON, Chapter 4, except to quote that

$$F_N(z) = \prod_{n=-N}^{N} (1 - z/\lambda_n)$$

for $|\lambda_n - n| \leq \rho < 1/4$, $\lambda_o \neq 0$, converges to an entire function $F(z)$ of exponential type π and $\int_{-\infty}^{\infty} (1 + x^2)^{-1} |F(x)|^2 \, dx < \infty$, so $F(z)/(z - \lambda_n)$ is in (PW) for each n, and consequently $f_n(z) = F(z)/(z - \lambda_n) F'(\lambda_n))$. Putting this into (1) gives

$$f(z) = F(z) \sum \frac{f(\lambda_n)}{(z - \lambda_n) F'(\lambda_n)} .$$

6.4.2.5 Completeness of the Sturm-Liouville eigenfunctions. A good example of a perturbed basis arises in connection with the eigenvalue problem for the Sturm-Liouville equation (cf. INCE, TRICOMI_2) on the interval $[0, 1]$. Corresponding to four different kinds of boundary conditions, we obtain normalized eigenfunctions $\{\varphi_n\}_{n=0}^{\infty}$ with the asymptotic behaviour

$$2^{-1/2} \varphi_n(x) = \begin{cases} \cos n\pi x + O(n^{-1}) \\ \cos (n + \frac{1}{2}) \pi x + O(n^{-1}) \\ \sin (n + \frac{1}{2}) \pi x + O(n^{-1}) \\ \sin (n + 1) \pi x + O(n^{-1}) \end{cases}$$

where the O is uniform in x as $n \to \infty$, (TRICOMI_2, p. 135). Now, this information, and the strong linear independence of the φ_n (i.e. no φ_j is in the closed space spanned by the remaining ones; in turn a consequence of the orthogonality of the φ_n) suffices to prove their completeness in $L^2(0, 1)$. Indeed, in each case we have a relation of the type $\varphi_n = \psi_n + O(1/n)$, where ψ_n is a known orthonormal basis for $L^2(0, 1)$. (In the second and third cases this is not quite obvious, and the reader should verify the completeness of $\{\psi_n\}$, e.g. with the aid of 6.4.1.1.) Now the following corollary to 6.4.2.2 immediately implies the completeness of the φ_n.

<u>6.4.2.6 Theorem</u>. <u>Let</u> $\{y_n\}_{n=1}^{\infty}$ <u>be an orthonormal basis for the Hilbert space</u> H, $\{x_n\}_{n=1}^{\infty}$ <u>strongly linearly independent elements of</u> H, <u>and suppose</u>

$$\sum_{n=1}^{\infty} \|x_n - y_n\|^2 < \infty .$$

<u>Then</u> $\{x_n\}_{n=1}^{\infty}$ <u>spans</u> H.

<u>Proof</u>. Choose N so large that $\sum_{n=N+1}^{\infty} \|x_n - y_n\|^2 < 1$. Then, since the system $\{0, \ldots 0, x_{N+1} - y_{N+1}, x_{N+2} - y_{N+2}, \ldots\}$ is a Bessel system with constant less than 1, it follows that $\{y_1, \ldots y_N, x_{N+1}, x_{N+2}, \ldots\}$ spans H and so the span S of $\{x_{N+1}, x_{N+2}, \ldots\}$ has deficiency at most N in H; since the adjunction of each vector $x_1, x_2, \ldots x_N$ to S successively diminishes the deficiency by 1, the result follows. ◇

We may remark that we have included the Sturm-Liouville example only to illustrate a certain technique - the "correct" proof of the completeness of the Sturm-Liouville eigenfunctions is that based upon the spectral theory of compact operators (or, "Hilbert-Schmidt Theorem" cf. $TRICOMI_2$).

Chapter 7. Minimal Extrapolation of Fourier Transforms

7.1 Introduction. As an application of the theory of best L^1 approximation, we now consider a problem in Fourier analysis, apparently first considered by Beurling. Not only has this problem considerable intrinsic interest; special cases of it play important roles (as we shall see) in obtaining sharp estimates in various problems of Fourier analysis, approximation theory, solution of functional equations, etc.

7.2 Notations. We begin by introducing some notation for Fourier analysis in the Euclidean n-space R^n and its dual, written $\hat{R}^n$. We employ usual vector notations; as a rule t, u shall denote points of R^n, and x, y points of $\hat{R}^n$. By tx we denote $\sum_{i=1}^n t_i x_i$, and dt, dx denote Haar measure in R^n, $\hat{R}^n$ respectively (for the uninitiated reader who is puzzled as to what $\hat{R}^n$ is, it is just Euclidean n-space (conceived, in abstract harmonic analysis, as the "dual group" of R^n); even when working in a Euclidean space, it is conceptually very helpful to think of the domain of a function and that of its Fourier transform as distinct copies of that space).

By $M(R^n)$ we denote the set of bounded complex measures on R^n, considered as a commutative Banach algebra under convolution (written $*$), with total variation as norm and unit element δ. The ideal $M_a(R^n)$ of absolutely continuous (with respect to Haar measure) measures is of course identified with $L^1(R^n)$.

$B(\hat{R}^n)$ denotes the set of Fourier transforms

$$\hat{\sigma}(x) = \int e^{-ixt}\, d\sigma(t)$$

of elements of $M(R^n)$, and is then a Banach algebra under ordinary pointwise multiplication, with $\|\hat{\sigma}\|_{B(\hat{R}^n)}$ defined as $\|\sigma\|_{M(R^n)}$ (when confusion is not to be feared, we write merely $\|\hat{\sigma}\|_B$ or even $\|\hat{\sigma}\|$ to simplify notations). Similarly $A(\hat{R}^n)$ denotes the set of Fourier transforms of measures in M_a, or what is the same thing, of functions (unspecified integrations always understood to be over R^n)

$$\hat{f}(x) = \int f(t) e^{-ixt}\, dt$$

where $f \in L^1(R^n)$, and $\|\hat{f}\|_{A(\hat{R}^n)}$ is defined to be $\|f\|_{L^1(R^n)}$. We occasionally speak of the "inverse Fourier transform", to refer to the "formally correct" formula $f(t) = (2\pi)^{-n} \int \hat{f}(x)e^{itx}\,dx$, which is valid when f is "nice enough" and can be rendered valid for any $f \in L^1$, as we shall see later, when a suitable "summability factor" is introduced into the integral. We shall take for granted, here and whereever else needed in these lectures, the elements of Fourier analysis (and sometimes a bit more). Unfortunately, although many excellent works on Fourier series exist, there is a dearth of modern literature dealing with the Fourier integral, especially in R^n. Useful sources are TITCHMARSH, BOCHNER & CHANDRASEKHARAN, RUDIN$_2$, DONOGHUE, GOLDBERG, KATZNELSON (and, to appear, STEIN & WEISS). However, some material we shall need is not available in these sources, but can usually be dug out of the "abstract" treatises, e.g. RUDIN$_1$, REITER, HEWITT & ROSS.

7.3 The problem of Beurling - general aspects

7.3.1 Notations. For notational simplicity, we shall state the problems and theorems on R, but we emphasize that in this general section (it will be otherwise later when we discuss explicit solutions of special problems!) all the results are valid in R^n (and even in quite general l.c.a. groups). Let E denote a closed subset of $\hat{R}$, and φ a complex-valued bounded continuous function on E. We define

$$\alpha(\varphi) = \inf \|F\|_{A(\hat{R})} \; ; \quad F \in A(\hat{R}) \;, \quad F = \varphi \text{ on } E.$$

$$\beta(\varphi) = \inf \|F\|_{B(\hat{R})} \; ; \quad F \in B(\hat{R}) \;, \quad F = \varphi \text{ on } E.$$

Clearly $0 \leq \beta(\varphi) \leq \alpha(\varphi) \leq \infty$. We define $\alpha(\varphi)$ or $\beta(\varphi)$ to be ∞ if the class of admissible F (i.e. those that extrapolate φ to a function of class $A(\hat{R})$, $B(\hat{R})$ respectively on the whole space) is empty. The problem to determine $\beta(\varphi)$ was posed by BEURLING, as well as that of determining an extrapolating F which has the minimal norm $\beta(\varphi)$, in case one exists. This minimizing F, Beurling then called a minimal extrapolation ("prolongement minimal") of φ, which we shall abbreviate m.e. (The problem of the m.e. can also be interpreted as the study of the quotient Banach space of $B(\hat{R})$ modulo the subspace of elements which vanish on E.)

Let us agree to call an element of $B(\hat{R})$ simply a minimal extrapolation (relative to E) if it is a m.e. of its restriction to E, in other words, if it is "orthogonal" to that subspace of $B(\hat{R})$ (in the sense of Chapter 4) consisting of elements which vanish on E. Similarly, we will also sometimes speak of a minimal extrapolation for the class $A(\hat{R})$ (relative to E) with an obvious meaning.

Of course, this problem can be much generalized, replacing $B(\hat{R})$ by all sorts of other classes; we have in 6.2 briefly touched on the analogous problem in Hilbert function spaces. But just the class $B(\hat{R})$ is essential for numerous applications. Needless to say, the concrete cases when $\beta(\varphi)$ or the minimal extrapolation are known explicitly are few, and therefore much interest also attaches to estimates for $\beta(\varphi)$. All that is really obvious is $\beta(\varphi) \geq \sup_{x \in E} |\varphi(x)|$.

Most of the hitherto explicitly solved cases of the problem are contained in SZ.-NAGY$_1$, and are limited to one variable. (Below we shall prove Sz.-Nagy's theorem by a new method, which will allow us even to squeeze out a few new cases.) The explicitly solvable cases found by Sz.-Nagy are extremely useful ones, and yield short and decisive proofs of Bohr's inequality, the Favard-Ahiezer-Krein approximation theorem, and other useful results. Let us first get some perspective, however, with some general theorems.

7.3.2 Definition. A point x of E is metrically isolated if the intersection of E with some neighborhood of x has measure zero.

7.3.3 Theorem. If φ possesses at least one extrapolation in $B(\hat{R})$, and E has no metrically isolated points, φ possesses a minimal extrapolation.

Proof. Let $\{F_n\}$ be a minimizing sequence, $\|F_n\|_B \longrightarrow \beta(\varphi) < \infty$. Write $F_n = \hat{\sigma}_n$, where $\sigma_n \in M(R)$. Since $\|\sigma_n\|_M$ is bounded we may clearly assume, by passing to a subsequence, that $\{\sigma_n\}$ is weak* convergent to some $\sigma \in M(R)$, that is $\int v\, d\sigma_n \longrightarrow \int v\, d\sigma$ for every $v \in C(R)$ vanishing at infinity. (But for the restriction to vanish at infinity, we would be done, since we could take $v(t) = e^{-ixt}$, with $x \in E$, and conclude $\hat{\sigma}(x) = \lim \hat{\sigma}_n(x) = \varphi(x)$, $x \in E$, even without any restriction on E! This is not permissible, however, and even (as we shall see) gives a false result. So we proceed as follows.) Let $g \in L^1(\hat{R})$. Then, by Fubini,

$$\int \hat{\sigma}_n(x)g(x)dx = \int G(t)d\sigma_n(t) \longrightarrow \int G(t)d\sigma(t) = \int \hat{\sigma}(x)g(x)dx$$

where $G(t) = \int g(x)e^{-itx}\,dx$ vanishes at ∞ by the Riemann-Lebesgue theorem. We have thus shown: $\hat{\sigma}_n \longrightarrow \hat{\sigma}$ in the weak* topology of L^∞. We next show

$$(*)\ \hat{\sigma}(x) = \varphi(x) \quad , \text{ almost all } x \in E \ .$$

Indeed, let K be a measurable subset of E having finite measure, and denote by ψ_K its characteristic function. By the above, $\int \hat{\sigma}_n \psi_K\,dx \longrightarrow \int \hat{\sigma}\,\psi_K\,dx$, and since $\hat{\sigma}_n = \varphi$ on E we get $\int_K (\varphi - \hat{\sigma})dx = 0$. Since this is true for each $K \subset E$ of finite measure, (*) is established. Thus, $\varphi - \hat{\sigma}$, which is continuous on E, vanishes on E outside a certain subset E_0 of measure zero. But, the hypothesis that E has no metrically isolated points implies that $E \setminus E_0$ is dense in E. Hence $\varphi - \hat{\sigma}$ vanishes on all of E, and since $\|\sigma\|_M \leq \beta(\varphi)$, $\hat{\sigma}$ is the desired minimal extrapolation of φ. ◇

7.3.4 Remarks. a) The previous theorem gives no information about "thin sets", e.g. sets of measure zero. Carl Herz has pointed out to me that for certain perfect sets of measure zero one can still assert the existence of a m.e. Indeed, the argument used in the above proof shows: if there exists a family N of bounded measures ν supported by E satisfying (i) The Fourier transform of each $\nu \in N$ vanishes at infinity, (ii) $\psi \in C(E)$ and bounded, and $\int \psi\,d\nu = 0$ for all $\nu \in N$ implies $\psi = 0$ then any $\varphi \in C(E)$ possessing an extrapolation in $B(\hat{R})$ possesses a m.e.

b) Concerning uniqueness of m.e., we refer the reader to DOMAR$_1$. There it is shown that the m.e. is unique when E is a half line, and half lines are the only sets with this property (i.e. that no function in $B(\hat{R})|_E$ has two distinct m.e.) On other sets, uniqueness may nevertheless hold for special subclasses of $B(\hat{R})|_E$, e.g. when E is the complement of (-1, 1) and φ decays fairly rapidly at infinity (loc. cit., p. 26). (The uniqueness when E is a half-line can be formulated thus, that the boundary functions of the Hardy space H^1 of the upper half-plane contain a unique closest element to each $f \in L^1(R)$. The analogous proposition for the circle had been proved earlier by DOOB.)

7.3.5 Exercises. a) Construct a compact set E and a function in $B(\hat{R})$ whose restriction to E has no m.e.

b) Prove that if $\varphi \in C(E)$ is the restriction of some $\hat{\sigma} \in B(\hat{R})$,

E is compact, and $\epsilon > 0$, then φ is the restriction of an element of $A(\hat{R})$ of norm at most $\|\sigma\| + \epsilon$ (and hence $\alpha(\varphi) = \beta(\varphi)$). (Hint: multiply $\hat{\sigma}$ by a suitably chosen function of compact support which equals 1 on E.)

c) Let K denote a circular arc in R^2, and μ that element of $M(R^2)$ which is arc length measure along K. Prove that $\hat{\mu}$ vanishes at infinity. Use this to deduce that every $\varphi \in C(E)$ with $\beta(\varphi) < \infty$ possesses a m.e., where E is the unit circumference. Prove the analogous result when E is the unit sphere in Euclidean n-space. (This result can actually be extended to any piece of a sufficiently smooth hypersurface in R^n, whose "curvature" is bounded away from zero.)

The following theorem was shown to me by Y. Domar, and it is implicit in DOMAR_1.

<u>7.3.6 Theorem.</u> <u>If there exists at least one absolutely continuous measure</u> σ <u>such that</u> $\hat{\sigma}|_E = \varphi$, <u>then</u> $\alpha(\varphi) = \beta(\varphi)$.

<u>Proof</u>. Let μ be any element of $M(R)$ such that $\hat{\mu}|_E = \varphi$. It is clearly sufficient to show that, given $\epsilon > 0$, there is $\tau \in M_a(R)$ such that $\hat{\tau}|_E = \varphi$ and $\|\tau\| < \|\mu\| + \epsilon$. Let now $\sigma_n \in M_a(R)$, $n = 1, 2, \ldots$ be an approximate identity for $M_a(R)$, i.e. $\|\sigma_n\| \leqslant 1$ and $\|(\sigma_n * \nu) - \nu\| \longrightarrow 0$ for each $\nu \in M_a(R)$. (For example, we can take $d\sigma_n = (n/2)e^{-n|t|}\,dt$, cf. SHAPIRO_1, Chapter 2.) Now by hypothesis there is some $\sigma \in M_a(R)$ satisfying $\hat{\sigma}|_E = \varphi$. Define $\tau_n = \sigma + (\mu - \sigma) * \sigma_n$. Then $\hat{\tau}_n = \hat{\sigma} = \varphi$ on E, and since $M_a(R)$ is an ideal in $M(R)$, $\tau_n \in M_a(R)$. Finally,

$$\|\tau_n\| \leqslant \|\sigma - (\sigma * \sigma_n)\| + \|\mu\|\,\|\sigma_n\| < \|\mu\| + \epsilon$$

for large n, and the theorem is proved. $\diamond$

An important consequence of this theorem is that an element of $A(\hat{R})$ which is a m.e. <u>for the class</u> $A(\hat{R})$ of its restriction to E, is necessarily a m.e. also for the larger class $B(\hat{R})$.

For the following theorem, taken from DOMAR_1, the following definition is needed. For a closed set $E \subset \hat{R}$, $L^1_E(R)$ denotes the set of $f \in L^1(R)$ such that $\hat{f}$ vanishes on E. By $L^\infty_E(R)$ we denote the set of $g \in L^\infty(R)$ such that $\int f(t)g(t)dt = 0$ for every $f \in L^1_E(R)$.

7.3.7 Theorem. Let E be a closed subset of $\hat{R}$, and $F = \hat{f}$ an element of $A(\hat{R})$. If there exists $\psi \in L_E^\infty(R)$ satisfying $\|\psi\|_\infty = 1$ and

$$(1) \qquad \psi(t)f(t) = |f(t)| \qquad \text{a.e.}$$

then F is a minimal extrapolation (in the class $B(\hat{R})$) of $F|_E$. Conversely, if $\|F\|_A = \alpha(F|_E)$ there exists $\psi \in L_E^\infty(R)$ of norm 1 satisfying (1).

Proof. Suppose first that ψ with the stated properties exists. By virtue of results in Chapter 4, this implies that $\|f + g\|_1 \geq \|f\|_1$ for every $g \in L_E^1(R)$, which means the same as $\|F + G\|_A \geq \|F\|_A$ for every $G \in A(\hat{R})$ vanishing on E, hence $\|F\|_A = \alpha(F|_E)$. By Theorem 7.3.6, $\alpha(F|_E) = \beta(F|_E)$, and the first part of the theorem is proved.

Conversely, if $\|F\|_A = \alpha(F|_E)$, that is, F is orthogonal (in $A(\hat{R})$) to the closed subspace of $A(\hat{R})$ consisting of the elements which vanish on E, or what is the same thing, f is orthogonal (in $L^1(R)$) to $L_E^1(R)$, we know from Chapter 4 that there exists $\psi \in L^\infty(R)$ of norm one which satisfies (1) and $\int \psi(t)g(t)dt = 0$ for all $g \in L_E^1(R)$, that is, $\psi \in L_E^\infty(R)$. ◇

In the following theorem, a consequence of the preceding, the notion "set of spectral synthesis" from harmonic analysis occurs (cf. KATZNELSON, RUDIN_1, EDWARDS). Although this notion is still largely unclarified, the following criteria are known: a closed set whose boundary is countable, and also (in higher dimensions) a polyhedral set (RUDIN_1, p. 169) are sets of spectral synthesis; however, some "nice" sets, such as the unit sphere in R^3, are not.

We also require the notion of spectrum of a function in L^∞: by this we mean the support of its Fourier transform (understood in the distributional sense) or, what comes to the same thing, a point $x \in \hat{R}$ is in the spectrum of $\psi \in L^\infty(R)$ if and only if, given any neighborhood N of x, there is some $f \in L^1(R)$ whose Fourier transform vanishes outside N, such that $\int f(t)\psi(-t)dt \neq 0$. The connection of these notions with the preceding theorem is this: If $\psi \in L_E^\infty(R)$, the spectrum of ψ is a subset of $-E = \{x: -x \in E\}$. Conversely, if E is a set of spectral synthesis, and $\psi \in L^\infty(R)$ has as its spectrum a subset of $-E$, then $\psi \in L_E^\infty(R)$. This enables us to assert:

7.3.8 Theorem. Let E be a closed subset of $\hat{R}$, $\varphi \in C(E)$ and $f \in L^1(R)$. If $\hat{f}|_E = \varphi$, E is a set of spectral synthesis and sgn f has its spectrum in E, then $\|f\|_1 = \alpha(\varphi) = \beta(\varphi)$ (i.e., $\hat{f}$ is a m.e. of φ).

Conversely, if $f \in L^1(R)$, $Z(f) = \{t: f(t) = 0\}$ has measure zero and $\|f\|_1 = \alpha(\hat{f}|_E)$, then sgn f has its spectrum in E.

The point of this is that in looking for a m.e., at least if we have reason to suspect or hope that a solution exists of the form $\hat{f}$, where f is in $L^1(R)$ and does not vanish on any set of positive measure, we can first try to find a unimodular function ψ whose spectrum lies in E. Such ψ play a role here precisely analogous to the role of extremal signatures in the theory of best uniform approximation. Indeed, each $f \in L^1(R)$ whose signum is one of these ψ has an extremal property, namely that $\hat{f}$ is a m.e. in the class $A(\hat{R})$ (and even in $B(\hat{R})$) of its restriction to E. Of course, in a given problem we must try to cook things up so that $\hat{f}$ coincides on E with the given function φ. Needless to say this is quite a large order, and we cannot expect in general to find an explicit solution, any more than we can in problems of best approximation generally. Already the problem of characterizing the "extremal signatures", that is the unimodular functions with spectrum lying in E, is quite intractable. In the Sz.-Nagy case, when E is the complement of the open unit interval, we must thus look for ψ with no spectrum in $(-1, 1)$, functions with a "spectral gap" in the terminology of B.F. Logan, Jr. who has made a penetrating study of these functions (some of his results shall be described below).

7.3.9 A digression on the philosophy of extremal problems

We might as well remark here that, strictly speaking, all the sophisticated considerations introduced thus far in this chapter are not necessary in order to demonstrate the known results concerning explicitly solvable cases of m.e. The point, which deserves emphasis, has arisen before (e.g. in connection with the Landau problem in Chapter 5), and is germane to the whole area of best approximation. Namely, in a typical case, one derives necessary and sufficient conditions for a b.a. Now, as a rule, the sufficiency of the conditions is rather trivial to prove, whereas the necessity lies much deeper. Moreover, the conditions in question (usually arrived at by a "variational" argument) are seldom of such an explicit form as to allow an explicit solution of the problem. Therefore, the only practical value of the necessity part of the theorem is, as a rule, moral support: it tells us that, by looking for an extremal among the functions satisfying such and such conditions (conditions which trivially suffice to make a function extremal) we are really not overlooking any possibilities, that the search in principle must succeed. Now, in practice, the way analysts most often demonstrate extremal properties is by proving directly some (often quite unnatural-looking) inequality, and then exhibiting, lo and behold!, a function for which the equality holds. However we emphasize for the benefit of students, who are often mystified by proofs of this character, that ordinary mortals are not likely to find such proofs without the benefit of a guiding principle (in the present context, the at least intuitive knowledge of necessary conditions, arrived at by variational considerations) although it might well be logically possible to present the demonstration with no reference to this principle. In numerical work (where it cannot be a question of finding an "explicit" solution, in the old-fashioned sense) general existence, uniqueness and characterization theorems play an important role in guiding the construction of algorithms, proving their convergence, deriving error estimates, etc.

7.3.10 Positive definite functions. We recall that a positive definite function on $\hat{R}$ is the Fourier transform of a bounded positive measure on R (this is not the usual definition, but in view of a theorem of Bochner (see KATZNELSON, p. 137) it comes to the same thing; F being positive definite is customarily defined (equivalently) to mean that the function $(x, y) \longrightarrow F(x - y)$ is positive definite, (or, "of positive type"), in the sense of Chapter 6).

7.3.10.1 Theorem. *Let* E *be a closed subset of* $\hat{R}$ *and* $\varphi \in C(E)$. *Then*

$$\beta(\varphi) \geqslant \sup_{x\in E} |\varphi(x)| \tag{1}$$

Moreover, if $|\varphi(x_0)| = \beta(\varphi)$ *for some* $x_0 \in E$ *and a minimal extrapolation exists, this m.e. is a constant multiple of a translate of a positive definite function.*

Proof. (1) is obvious. Assuming the other hypotheses, and making a translation which takes x_0 to the origin, we see that $\hat{\sigma}$, the m.e., satisfies $|\varphi(0)| = |\int d\sigma| = \|\sigma\|$, implying that σ is a constant multiple of a positive measure. ◇

Observe also that a positive definite function, or translate thereof, always attains a maximum, and this maximum value is its $B(\hat{R})$ norm.

As an application, let us prove the Bernstein inequality (cf. Chapter 3) by another method. Suppose f is a trigonometric polynomial of order n, $f(t) = \sum_{-n}^{n} c_k e^{ikt}$, then

$$f'(t) = \sum_{-n}^{n} ik\, c_k\, e^{ikt} = i\int f(t-u)\,d\sigma(u) \tag{2}$$

where σ is any measure on R satisfying $\hat{\sigma}(k) = k$, $k \in E = \{-n, -n+1, \ldots, n\}$. Hence $\|f'\|_\infty \leqslant \|f\|_\infty \|\sigma\|_M$, and since we have freedom to choose σ, $\|f'\|_\infty \leqslant \|f\|_\infty\, \beta(\varphi)$ where $\varphi(x) = x$, $x \in E$. Here it is quite easy to find a m.e. Indeed, once we suspect that $\beta(\varphi) = n$, we try to extrapolate φ to a translate of a positive definite function. A natural candidate is the <u>periodic</u> function $\hat{\sigma}(x)$ equal to x on $[-n, n]$ and extended so as to be symmetric about n, and have period 4n. In other words, we try $\hat{\sigma}(x) = (2n/\pi)\, F((\pi x/2n) - (\pi/2))$ where F is the standard "saw-tooth" function of period 2π, even, and equal to $(\pi/2) - x$ for $0 \leqslant x \leqslant \pi$. Clearly, if F is positive definite, σ is a translate of a p.d. function and its norm must be $\max|\hat{\sigma}(x)| = n$, yielding the Bernstein inequality. Now, F is the Fourier transform of a measure supported on the integers, the mass at $t = k$ being

$$a_k = (2\pi)^{-1}\int_{-\pi}^{\pi} F(x)e^{-ikx}\,dx = \begin{cases} 0, & k = 0 \\ (\pi k^2)^{-1}(1-\cos k\pi), & k \neq 0 \end{cases}$$

and since $a_k \geqslant 0$, F is positive definite, and we are done.

It is worth while to look a little more closely at the form which (2) takes when we choose for σ the extremal measure. We have

$$i\hat{\sigma}(x) = 4n\pi^{-2} \sum_{k \text{ odd}} (-1)^{\frac{k-1}{2}}\, k^{-2}\, e^{ik\pi x/2n}$$

so that σ is the discrete measure with mass $4n(-1)^{(k-1)/2}(\pi k)^{-2}$ at $t = k\pi/2n$, $k = \pm 1, \pm 3, \ldots$, and (2) becomes

$$f'(t) = 4n\pi^{-2} \sum_{k \text{ odd}} (-1)^{(k-1)/2} k^{-2} f(t - (k\pi/2n)) \tag{3}$$

(the sum being over <u>all</u> odd integers, positive and negative). Formula (3) is valid for every trigonometric polynomial f of degree at most n, and it "puts in evidence" the Bernstein inequality, since the sum of the absolute values of the coefficients appearing on the right side is n. This remarkable interpolation formula for the functional $f'(t)$ is an essentially different one than that we discussed in Chapter 3. Just as that formula was a Riesz representation for the (Hahn-Banach extension to $C(T)$ of the) derivative, when the trigonometric polynomials are embedded in $C(T)$, so the present one can be viewed as a Riesz representation corresponding to the embedding of the trigonometric polynomials in $C(bR)$, where bR denotes the Bohr compactification of R (KATZNELSON, p. 192); in principle, the representing measure might have been carried partly on the "ideal" part of bR, although in fact this turned out not to be the case.

It is easy to see that the above analysis is valid for any continuous $f \in L^\infty(R)$ whose spectrum lies in $[-n, n]$ (where now there is no reason any more to suppose n an integer), or (what can be shown to be the same thing) the restriction to the real axis of an entire function of exponential type not exceeding n (cf. ZYGMUND, vol. II, p. 276). Finally, we remark that (3) yields also the L^p version of the Bernstein inequality, $\|f'\|_p \leq n\|f\|_p$ for $1 \leq p \leq \infty$.

<u>Remark</u>. The reader interested in pursuing further the Bernstein inequality and its generalizations, related interpolation formulae, etc., from the standpoint of complex analysis will find a wealth of material in Chapter 11 of BOAS and further works referred to there.

<u>Exercises</u>. a) Find a minimal extrapolation of $\varphi(x) = e^{|x|}$ from $E = [-c, c]$.

b) Find a m.e. of $\varphi(x) = e^{-x}$ from $E = [0, 1]$. Generalize the result of this and of the preceding exercise to convex functions.

c) Use the above method to find all cases of equality in the Bernstein inequality (do also the L^p case, treating both trigonometric polynomials in the $L^p(-\pi, \pi)$ norm and functions of exponential type in the $L^p(R)$ norm).

d) Prove that if f is entire, of exponential type at most π, and bounded on the real axis

$$f'(z) = (4/\pi) \sum_{n=-\infty}^{\infty} \frac{(-1)^n f(z + n + \frac{1}{2})}{(2n+1)^2}$$

for all complex z, the convergence being uniform on each horizontal strip.

e) Let $1 < c \leq 3/2$, define E to be the union of the intervals $(-\infty, -c]$, $[-1, 1]$, and $[c, +\infty)$, and consider the function $\varphi = \varphi_c$ on E defined as 1 on $[-1,1]$ and 0 on the remaining two intervals. Prove that $A_1 \log (1/(c-1)) \leq \alpha(\varphi) \leq$ $\leq A_2 \log (1/(c-1))$, where A_i denote absolute constants. (Hint: for the upper bound, extend φ linearly into $[-c, -1]$ and $[1, c]$ and observe that the resulting "trapezoid function" F_c is the Fourier transform of $f_c(t) \underset{\text{def.}}{=} (1/\pi(c-1))(\cos t - \cos ct)/t^2$.

An elegant way to compute $\|f_c\|_1$ is to observe that

$$|f_c(t)| = f_c(t) \operatorname{sgn} \sin ((c+1)/2)t \operatorname{sgn} \sin ((c-1)/2)t$$

and, oblivious to the pitfalls of conditionally convergent series, plug in Fourier expansions of the two sgn sin expressions and integrate termwise. For the lower bound, show how to express the nth partial sum of the Fourier series of an arbitrary continuous periodic function with the aid of the function f_c, where $n = [1/(c-1)]$.)

Remark. To our knowledge, the exact value of $\alpha(\varphi)$ has not been determined. Kernels whose Fourier transforms extrapolate the above function φ_c are of importance in the study of trigonometric series which involve arbitrary frequencies, the dilations of these kernels taking over the role of the Dirichlet kernel.

f) Obtain estimates, in terms of c, n for the problem analogous to e) in R^n (i.e. take $\varphi = 1$ inside the unit ball, and $= 0$ outside the ball of radius c).

7.3.11 Extrapolating the "tail" of a Fourier transform. For later purposes we shall require the following result, which is useful in many questions of Fourier analysis.

7.3.11.1 Theorem. Let $F \in A(\hat{R})$, $E_\lambda = \{x: |x| \geq \lambda\}$, and denote by φ_λ the restriction of F to E_λ. Then $\lim_{\lambda \to \infty} \alpha(\varphi_\lambda) = 0$.

Proof. Let $F = \hat{f}$, $f \in L^1(R)$, and let k be an integrable function on R such that $\int k\, dt = 1$ and $K = \hat{k}$ has compact support (e.g. take $k(t) = \pi^{-1}(\sin t/t)^2$). Clearly it is enough to show that the $A(\hat{R})$ norm of the function $F(x)(1 - \hat{k}(\epsilon x))$ tends to zero when $\epsilon \to 0$. But this function is the Fourier transform of $f - (f * k_{(\epsilon)})$ where $k_{(\epsilon)}(t) = \epsilon^{-1} k(t/\epsilon)$ is an "approximate identity" so that $\|f - (f * k_{(\epsilon)})\|_1 \to 0$ as $\epsilon \to 0$ (cf. SHAPIRO$_1$, Chapter 2). ◇

In a similar manner one can show, if $F \in A(\hat{R})$ and $F(x_0) = 0$, and ψ_ϵ denotes the restriction of F to $\{x: |x - x_0| \leq \epsilon\}$, that $\lim_{\epsilon \to 0} \alpha(\psi_\epsilon) = 0$ (exercise!)

As an application of Theorem 7.3.11.1 we have (we state the result specifically in R^n, because we will need it later):

7.3.11.2 Corollary. If $\sigma \in M(R^n)$, $F \in A(\hat{R^n})$ there exists $\tau \in M(R^n)$ such that $\hat{\tau}(x)(1 - F(x)) = \hat{\sigma}(x)$ in a neighborhood of infinity.

Proof. We can find $G \in A(\hat{R^n})$ of norm at most 1/2 which coincides with F in a neighborhood of ∞, by the preceding theorem. Therefore $1 - G$ is invertible in the Banach algebra $B(\hat{R^n})$ (KATZNELSON, p. 198), and so $\hat{\tau}(x) = \hat{\sigma}(x)(1 - G(x))^{-1}$ defines a measure τ which satisfies the requirement. ◇

It will be important for our later work to consider divisibility problems in $B(\hat{R^n})$. Given any class D of functions on a topological space X, we say of a pair of elements f, g of D that f divides g (in D) iff $fh = g$ for some $h \in D$, and that f divides g (in D) at the point $x_0 \in X$ iff there exists $h \in D$ such that $f(x)h(x) = g(x)$ for all x in a neighborhood of x_0. In like manner, we say that a function f defined on X belongs to D at x_0 iff there exists $g \in D$ such that $g(x) = f(x)$ for all x in a neighborhood of x_0. An important basic result of Fourier analysis is that $B(\hat{R^n})$ is a local algebra, in this sense: a function F defined on $\hat{R^n}$ which belongs to $B(\hat{R^n})$ at each point of $\hat{R^n}$ and also at infinity (that is, there is an element of $B(\hat{R^n})$ which coinsides with F in some neighborhood of infinity), is an element of $B(\hat{R}^n)$. We also recall to the reader the famous <u>Wiener lemma</u>, that for $F \in B(\hat{R}^n)$, $1/F$ belongs to $B(\hat{R})$ at any point x_0 where $F(x_0) \neq 0$. Observe that $A(\hat{R}^n)$ and $B(\hat{R}^n)$ are <u>locally</u> the same except at infinity. We can now state

<u>7.3.11.3 Corollary.</u> *If $F \in A(\hat{R}^n)$, $G \in B(\hat{R}^n)$ and $1 - F$ divides G (in $B(\hat{R}^n)$) at each point $x \in \hat{R}^n$ where $F(x) = 1$, then $1 - F$ divides G (in $B(\hat{R}^n)$).*

<u>Proof.</u> It is enough to check that $1 - F$ divides G at each point of $\hat{R}^n$, and at infinity. The latter is taken care of by the preceding corollary, points where $F(x) = 1$ by the hypothesis, and points where $F(x) \neq 1$ by Wiener's lemma. $\diamond$

<u>Exercises.</u> a) Let $k(t) = \pi^{-1}(\sin t/t)^2$, $h(t) = (\pi(1 + t^2))^{-1}$. Prove that $\delta - k dt$ and $\delta - h dt$ mutually divide one another in the convolution measure algebra $M(R)$.

b) (Continuation.) Prove that the quotients have again the form $\delta - p\, dt$, with $p \in L^1(R)$.

c) Let $F_1, \dots F_n$ be elements of $A(\hat{R})$, $G_n = 1 - F_n$, and suppose $\sum_{i=1}^{n} |G_i(x)| \geq 1$ for all $x \in \hat{R}$. Prove that there exist elements $H_i \in B(\hat{R})$ such that $\sum_{i=1}^{n} H_i(x)\, G_i(x) = 1$ for all x.

d) Let M_0 denote the closed subalgebra of M generated by the unit element δ and M_a (the absolutely continuous measures). Describe the maximal ideal space of M_0.

<u>7.4 Explicitly solvable cases of m.e.: Sz.-Nagy's theorem</u>

<u>7.4.1 General remarks.</u> Throughout this section we shall work on $\hat{R}$, E <u>shall always denote the complement of</u> $(-1, 1)$, and we suppose given a bounded function $\varphi \in C(E)$. We may as well suppose (since it is only for such φ that the method below works) that φ can be extended to an element of $A(\hat{R})$. By theorem 7.3.3 φ possesses a m.e., call it F, in $B(\hat{R})$. Then $F - \varphi$ is in $B(\hat{R})$ and has compact support, hence (exercise!) it is in $A(\hat{R})$, so that also $F \in A(\hat{R})$. (Incidentally, since by the F. and M. Riesz theorem an element of $B(\hat{R})$ which vanishes on a half-line lies in $A(\hat{R})$, the same argument shows that if φ is the restriction to <u>any</u> closed set E which contains a

half line and has no metrically isolated points, of an element of $A(\hat{R})$, then φ possesses a m.e. in the class $A(\hat{R})$.)

From the general remarks following Theorem 7.3.8 we know that good candidates for m.e. are functions $F = \hat{f}$, $f \in L^1(R)$ <u>such that</u> sgn f <u>has no spectrum in</u> $(-1, 1)$. A very simple class of bounded functions with no spectrum in $(-1, 1)$ is <u>functions with period</u> 2π <u>and mean zero</u>. Indeed, the (distributional) Fourier transform of such a function ψ is a distribution supported by the integers, and 0 will be absent from the support if ψ has mean value zero. And within this class, the simplest members are

(1) $$\psi_c(t) = \operatorname{sgn} \cos t$$

and its translate

(2) $$\psi_s(t) = \operatorname{sgn} \sin t$$

Thus, if we take any real-valued function $f \in L^1(R)$ which changes sign precisely where cos t does, or precisely where sin t does, we know by the preceding theory that $\hat{f}$ will be a m.e. of its restriction to E. The question is whether f can be chosen so that this restriction turns out to be something in which we are interested. Well, what would we consider "interesting"? Since in this chapter we have given a systematic, rather than historical development, it is necessary to anticipate by a little the applications, and say that $\varphi(x) = x^{-r}$, $r = 1, 2, \ldots$ are very interesting for applications, and likewise $\varphi(x) = e^{-a|x|}$ and $(\cosh ax)^{-1}$. The theorem (of Sz.-Nagy) at which we are aiming shows that, quite remarkably, all of these cases, in fact all <u>even</u> and all <u>odd</u> φ with suitable monotonicity properties, can be handled by use of the simple signum functions (1), (2) respectively.

7.4.2 Before proceeding, let us fix some notation. We write

$$(C) = \left\{ f \in L^1(R) \; : \; f(t) \cos t \geq 0 \quad \text{a.e.} \right\}$$

$$(S) = \left\{ f \in L^1(R) \; : \; f(t) \sin t \geq 0 \quad \text{a.e.} \right\} .$$

These are closed convex cones in $L^1(R)$. We write (C)^, (S)^ to denote the images of these classes under Fourier transformation. By P we denote the set of positive definite functions on $\hat{R}$, and by P_a the subclass of P consisting of the Fourier transforms of non-negative integrable functions. By P^* we denote the set of functions on $\hat{R}$ which are continuous, even, and convex and decrease to zero on $[0, \infty)$. It is well known that $P^* \subset P_a$ (TITCHMARSH p. 170, DONOGHUE p. 187). In the following discussion we find it useful in places to employ the concepts of distribution theory, although that could easily be circumvented.

The following lemmas are due to SZ.-NAGY$_1$ (the proofs are implicit in what we have already said, apart from the formulae for $\|F\|_A$).

7.4.3 a) Lemma. Every $F \in (C)^\wedge$ is a minimal extrapolation relative to E. Moreover, if F is even,

(1)
$$\|F\|_A = \frac{4}{\pi} \left|\frac{F(1)}{1} - \frac{F(3)}{3} + \frac{F(5)}{5} - \dots\right|$$

b) Lemma. Every $F \in (S)^\wedge$ is a minimal extrapolation relative to E. Moreover, if F is odd

(2)
$$\|F\|_A = \frac{4}{\pi} \left|\frac{F(1)}{1} + \frac{F(3)}{3} + \frac{F(5)}{5} + \dots\right|$$

Proofs. The proofs of (1) and (2) are based on the Fourier expansions

(3)
$$\operatorname{sgn} \cos t = \frac{4}{\pi} \sum_{n=0}^{\infty} (-1)^n \frac{\cos(2n+1)t}{2n+1}$$

(4)
$$\operatorname{sgn} \sin t = \frac{4}{\pi} \sum_{n=0}^{\infty} \frac{\sin(2n+1)t}{2n+1}$$

We treat only the odd case (the other being handled in exactly the same way). Writing $F = \hat{f}$, we have

(5)
$$\|F\|_A = \int |f(t)|dt \geq \left|\int f(t) \operatorname{sgn} \sin t \, dt\right| = \left|(4/\pi) \sum_{\substack{m=1 \\ m \text{ odd}}}^{\infty} (\hat{f}(-m) - \hat{f}(m))/2im\right|$$
$$= (4/\pi) \left| \sum_{n=0}^{\infty} F(2n+1)/2n+1 \right| \qquad \diamond$$

We shall not follow completely Sz.-Nagy's method; instead we base our argument on positive definite functions, the relation of which to m.e. relative to E is contained in the next lemmas.

7.4.4 a) Lemma. $F \in (C)^\wedge$ if and only if $G(x) = F(x+1) + F(x-1)$ is positive definite. In particular, if this occurs, F is a minimal extrapolation relative to E.

b) Lemma. $iF \in (S)^\wedge$ if and only if $H(x) = F(x-1) - F(x+1)$ is positive definite. In particular, if this occurs, F is a minimal extrapolation relative to E.

Proofs. It suffices to consider the second lemma. If $F = \hat{f}$, then H is the Fourier transform of $f(t)(e^{it} - e^{-it}) = 2i \sin t\, f(t)$, which is real and non-negative if and only if $iF \in (S)^\wedge$. ◇

It is now easy to deduce a criterion for m.e. from which practical consequences can be drawn easily.

7.4.5 Theorem. An element F of $A(\hat{R})$ belongs to $(C)^\wedge$ if and only if the difference equation

$$(1) \qquad K(x+1) + K(x-1) = F(x)$$

has at least one solution which is $O(|x|)$ at infinity, and such that $k(t) \cos^2 t$ is a non-negative integrable function on R. (Here k denotes the tempered distribution on R whose Fourier transform is K.) In particular, any F for which such K can be found is a m.e. relative to E.

7.4.6 Theorem. An element F of $A(\hat{R})$ belongs (after multiplication by i) to $(S)^\wedge$ if and only if the difference equation

$$(1) \qquad K(x-1) - K(x+1) = F(x)$$

has at least one solution which is $O(|x|)$ at infinity, and such that $k(t) \sin^2 t$ is a non-negative integrable function on R. (Here k denotes the tempered distribution on R whose Fourier transform is K.) In particular, any F for which such K can be found is a m.e. relative to E.

As immediate consequences of these theorems we have

7.4.7 Corollary. If either 7.4.5 (1) or 7.4.6 (1) has a solution K which is a positive definite function (or constant multiple thereof), F is a m.e. relative to E.

Proofs. Theorems 7.4.5 and 7.4.6 are proved in the same way (indeed, they can be formulated as special cases of a single more general theorem, as we'll see below; we have however separated them, for convenience in applications), and we confine attention to 7.4.5. Suppose first $F \in (C)^{\wedge}$. Since it is bounded, (1) obviously has a solution K that is $O(|x|)$ at infinity. By Fourier transformation, 7.4.5 (1) becomes

$$k(t)(e^{it} + e^{-it}) = f(t), \text{ where } \hat{f} = F.$$

Now, by assumption $f(t) \cos t \geq 0$ a.e., hence $2k(t) \cos^2 t$ equals the non-negative integrable function $f(t) \cos t$. Conversely, if 7.4.5 (1) has a solution with the stated properties, we deduce that $f(t) \cos t \geq 0$ a.e. ◇

Deduction of the Corollary. Suppose, say, 7.4.5 (1) has a positive definite solution K. Then k is a bounded positive measure, hence so is $2k \cos^2 t$. Since this equals $f \cos t$, the latter is an integrable function which is also a positive measure, hence as a function it is non-negative a.e. ◇

7.4.8 Example. The Corollary suggests a strategy for constructing m.e. as a superposition of two translates of a positive definite function. Suppose, to fix ideas, that we are given an even function φ on E (say, $\varphi(x) = x^{-2}$) and we wish to construct a m.e. In this example it is almost obvious from geometric considerations that there is a function K of class P^* (as defined in 7.4.2) such that $K(x + 1) + K(x - 1) = x^{-2}$ for $|x| \geq 1$ (draw a picture!). The function $F(x) = K(x + 1) + K(x - 1)$, the m.e. of x^{-2}, has then a graph resembling a suspension bridge, with cusps at $x = \pm 1$. From the difference equation we immediately get, for $x \geq 0$, a formula for this particular solution:

$$(1) \qquad K(x) = (x + 1)^{-2} - (x + 3)^{-2} + (x + 5)^{-2} - \dots$$

which is readily checked to be decreasing, convex, and tend to zero. Hence the solution K of class P^* is

$$K(x) = \sum_{n=0}^{\infty} (-1)^n (|x| + 2n + 1)^{-2}$$

and finally, the desired m.e. of x^{-2} is

$$(2) \qquad F(x) = \sum_{n=0}^{\infty} (-1)^n [(|x+1| + 2n + 1)^{-2} + (|x-1| + 2n + 1)^{-2}]$$

$$= \sum_{m=1}^{\infty} (-1)^{m-1} [(2m + x)^{-2} + (2m - x)^{-2}]$$

when x is in the "interesting" range, $[-1, 1]$. This is just the solution Sz.-Nagy had arrived at (as specialized to $\varphi(x) = x^{-2}$) by other considerations. In the present instance the task was specially simple, because (i) the "distinguished" solution of the difference equation which we were after, could be written down by inspection (formula (1)). In other cases, e.g. when φ is odd and $\sum_{n=1}^{\infty} \varphi(x + 2n + 1)$ diverges, as happens in the important example $\varphi(x) = x^{-1}$, solution of the difference equation is less trivial. (ii) The function K was positive definite, and that for the easily recognizable reason that it belonged to P^*. In other cases (that are none the less solvable by this method) this will not be so, e.g. for $\varphi(x) = x^{-1}$ the associated $K(x)$ is unbounded and we really need the "distributional" criterion of Theorem 7.4.6.

Exercises. a) Find a "closed" expression for $F(x)$ in $[-1, 1]$ in terms of trigonometric functions; sketch the graph.

b) F is infinitely differentiable away from the points $x = \pm 1$. Show that the second derivative, and all even order derivatives, are continuously extendible across the points ± 1, and these derived functions are all minimal extrapolations relative to E.

c) Compute $\alpha(\varphi)$ in the above example.

d) Let p denote a trigonometric polynomial of the form $p(t) = \sum_{k=m}^{n} c_k e^{ikt}$.

Show there exists a constant A such that $\|p\|_\infty \leqslant Am^{-2}\|p''\|_\infty$, and find the best possible value for this constant.

e) Carry out an analysis similar to the above for $\varphi(x) = |x|^{-1}$ and $\varphi(x) = x^{-3}$; to what inequalities for trigonometric polynomials do these results lead? Carry out a similar analysis for $\varphi(x) = x^{-1}$, solving the difference equation with the aid of gamma functions.

We require two more lemmas before the final results.

__7.4.9 Lemma__. __Let__ K __be a real-valued function on__ R, __even and twice differentiable on__ $(0, \infty)$, __satisfying__ $K''(x) \geqslant 0$ __for__ $x > 0$ __and__ $\lim_{x \to \infty} K'(x) = 0$. __Then__ K __satisfies the hypotheses of Theorem__ 7.4.6. (__Hence, if__ $K(x-1) - K(x+1)$ __is an element of__ $\hat{A}(R)$, __it is a m.e. relative to__ E.)

__Proof__. Denoting by D the operation of differentiation, in the sense of distribution theory, we readily verify that

$$D^2K = 2K'(0+)\,\delta + J$$

where J is the function defined for $x \neq 0$ by $J(x) = K''(|x|)$. The hypotheses imply $J \in L^1(R)$. Denoting by j, k the ("inverse") Fourier transforms of J, K we have therefore $-t^2k = 2K'(0+) + j$. From this we see that $k \sin^2 t$ equals the integrable function $(-2K'(0+) - j(t)) \sin^2 t/t^2$, and to complete the proof we must show that this function is non-negative, i.e. $|j(t)| \leqslant -2K'(0+)$. But

$$|j(t)| = 2\left|\int_0^\infty K''(x)\,e^{itx}\,dx\right| \leqslant 2\int_0^\infty K''(x)\,dx = -2K'(0+)$$

completing the proof. ◇

An example of K satisfying the hypotheses is $K(x) = -\log(1 + |x|)$. Our final lemma is a simple fact about difference equations.

__7.4.10 Lemma__. __Let__ ψ __denote a twice differentiable function on__ $(0, \infty)$ __which is positive, convex and decreasing to zero__. __For each__ $c > 0$ __the difference equation__

(1) $$K(x) - K(x + c) = \psi(x)$$

<u>has a solution</u> K <u>which satisfies on</u> $(0, \infty)$ <u>the conditions in the preceding lemma</u>.

<u>Proof</u>. We observe first that if

$$(2) \qquad \int_0^\infty \psi(x)\, dx < \infty$$

holds, the series $\psi(x) + \psi(x + c) + \psi(x + 2c) + \ldots$ converges absolutely and furnishes a solution of (1), and it is readily checked that this solution has the required properties. In the general case, since $-\psi'$ is positive and integrable on $(0, \infty)$, the sum of the absolutely convergent series

$$-\psi'(x) - \psi'(x + c) - \psi'(x + 2c) - \ldots = J(x)$$

is a positive decreasing function, hence $K(x) = -\int_0^x J(y)\, dy$ is decreasing and $K'(x) = -J(x)$ tends to zero as $x \longrightarrow \infty$. Moreover, $K''(x) = -J'(x) > 0$. Finally,

$$\frac{d}{dx}(K(x) - K(x + c) - \psi(x)) = -J(x) + J(x + c) - \psi'(x) = 0$$

showing that $K(x) - K(x + c) - \psi(x)$ is a constant, which must be zero as we see upon letting $x \longrightarrow \infty$. ◇

<u>7.4.11 Theorem</u> (<u>SZ.-NAGY$_1$</u>). <u>Let</u> φ <u>be a real-valued function defined on</u> $E = R \setminus \setminus (-1, 1)$, <u>even, vanishing at infinity, and "three times monotonic" on</u> $[1, \infty)$, <u>that is it is decreasing, convex, and possesses a second derivative which decreases</u>. <u>Then</u> φ <u>possesses a minimal extrapolation</u> $F \in A(\hat{R})$ <u>defined by</u>

$$(1) \qquad F(x) = \sum_{m=1}^\infty (-1)^{m-1} (\varphi(2m - x) + \varphi(2m + x)) , \quad |x| \leq 1$$

<u>and</u>

$$(2) \qquad \alpha(\varphi) = \beta(\varphi) = \|F\|_A = (4/\pi) \sum_{n=0}^\infty (-1)^n \varphi(2n + 1)/(2n + 1) .$$

<u>7.4.12 Theorem</u> (<u>SZ.-NAGY$_1$</u>). <u>Let</u> φ <u>be a real-valued function defined on</u> $E = R \setminus \setminus (-1, 1)$, <u>odd, vanishing at infinity, and on</u> $[1, \infty)$ <u>possessing a positive second derivative</u>. <u>Suppose moreover</u> $\int_1^\infty \varphi(x)/x\, dx < \infty$. <u>Then</u> φ <u>possesses a minimal extra-</u>

<u>polation</u> $F \in A(\hat{R})$ <u>defined by</u>

(1) $$F(x) = \sum_{m=1}^{\infty} (\varphi(2m - x) - \varphi(2m + x)) , \quad |x| \leqslant 1$$

<u>and</u>

(2) $$\alpha(\varphi) = \beta(\varphi) = \|F\|_A = (4/\pi) \sum_{n=0}^{\infty} \varphi(2n + 1)/(2n + 1) .$$

<u>Remark</u>. The condition $\int_1^\infty \varphi(x)\,dx < \infty$ is equivalent to the existence of <u>some</u> extrapolation in $A(\hat{R})$ (exercise!).

<u>Proofs</u>. We have already done most of the work. Consider first 7.4.11. The even function K whose value for $x > 0$ is

$$K(x) = \varphi(x + 1) - \varphi(x + 3) + \varphi(x + 5) - \dots$$

satisfies $K(x - 1) + K(x + 1) = \varphi(x)$, $x \in E$. The hypotheses on φ imply that $K \in P^*$, hence 7.4.7 tells us that $F(x) = K(x - 1) + K(x + 1)$ is a m.e. of φ, which leads to (1), and (2) follows from 7.4.3.

As for 7.4.12, we know by 7.4.10 that there exists K satisfying $K(x) - K(x + 2) = \varphi(x + 1)$ for $x > 0$, that is, $K(x - 1) - K(x + 1) = \varphi(x)$ for $x > 1$, and moreover (when extended by evenness to R) satisfying the hypotheses of 7.4.9. This gives that $F(x) = K(x - 1) - K(x + 1)$ is a m.e. of φ, and the rest follows the same pattern as the preceding proof. ◇

<u>7.4.13 Examples</u>

a) $\varphi(x) = x^{-s}$, $s > 0$ for $x \geqslant 1$, and φ <u>odd</u>. Theorem 7.4.12 applies, and $\alpha(\varphi) = \Lambda(s)$ where

(1) $$\Lambda(s) = (4/\pi) \sum_{n=0}^{\infty} (2n + 1)^{-s-1}$$

In particular, $\Lambda(1) = (4/\pi)(\pi^2/8) = \pi/2$. The m.e. can be computed from 7.4.12 (1), for example when $s = 1$ it is

(2) $$N_1(x) = \frac{1}{x} - \frac{\pi}{2} \cot \frac{\pi x}{2} , \quad |x| \leqslant 1$$

(cf. SZ.-NAGY$_1$; ESSEEN, p. 13 ff.).

b) $\varphi(x) = x^{-s}$, $s > 0$ for $x \geqslant 1$, and φ even. Theorem 7.4.11 applies, and $\alpha(\varphi) = \tilde{\lambda}(s)$ where

$$(3) \qquad \tilde{\lambda}(s) = (4/\pi) \sum_{n=0}^{\infty} (-1)^n (2n+1)^{-s-1}$$

c) $\varphi(x) = e^{-cx}$, $c > 0$ for $x \geqslant 1$, and φ even. (This is equivalent to the problem of m.e. of $e^{-|x|}$ relative to $R \setminus (-c,c)$.) Theorem 7.4.11 applies, and

$$\alpha(\varphi) = (4/\pi) \text{ arc tan } e^{-c}$$

and the m.e. equals $(e^{2c} + 1)^{-1} (e^{cx} + e^{-cx})$ for $|x| \leqslant 1$.

d) $\varphi(x) = e^{-cx}$, $c > 0$ for $x \geqslant 1$ and φ odd. Theorem 7.4.12 applies and

$$\alpha(\varphi) = (2/\pi) \log [(e^c + 1)/(e^c - 1)]$$

the m.e. being $(e^{2c} - 1)^{-1}(e^{cx} - e^{-cx})$ for $|x| \leqslant 1$.

e) $\varphi(x) = (\cosh ax)^{-1}$, $|x| \geqslant 1$. (Cf. AHIEZER$_1$, p. 215). Here 7.4.11 applies, at least if a exceeds a suitable constant A, and

$$\alpha(\varphi) = (4/\pi)(\text{sech } a - (\text{sech } 3a)/3 + \dots) < (4/\pi) \text{ sech } a .$$

7.4.14 On the effectiveness of Sz.-Nagy's criterion. The Sz.-Nagy criteria for membership in the classes $(C)^\wedge$ and $(S)^\wedge$, although handy and covering the cases of greatest "practical" interest, is far from being necessary, as is evident from our previous discussion. Indeed, for any positive definite function K, we know that $K(x + 1) + K(x - 1)$ is in $(C)^\wedge$, although this function need not be three times, nor for that matter even one time, monotonic on $x \geqslant 1$. It is easy to construct a real-valued positive definite function K, vanishing outside $(-1, 1)$ and not monotone decreasing on $[0, 1]$.

Exercise. Construct such a function. (Hint: start from the function equal to $(|x| - \frac{1}{2})^2$ on $[-\frac{1}{2}, \frac{1}{2}]$, and zero outside, and modify it by adding on two small triangular "hats" centered at $\pm 1/2$.)

A perhaps less artificial-looking example is simply to consider the even function φ, defined by

$$\varphi(x) = \begin{cases} (5 - 2x)/3 & , \quad 1 \leqslant x \leqslant 5/2 \\ 0 & , \quad x \geqslant 5/2 \end{cases}$$

This function is convex, but not <u>three</u> times monotonic (not even if we consider this in the extended sense, as Sz.-Nagy actually does, i.e. not <u>a priori</u> requiring differentiability but merely that the forward differences up to order three are positive). Thus, if we consider the m.e. problem for the <u>evenly</u> extended φ, 7.4.11 is not applicable. On the other hand, the method of superposition immediately gives us a m.e. of class $(C)^\wedge$. For, letting $K(x)$ denote the (positive definite!) "triangle function" centered at 0, of height 1 and semi-base 3/2, $K(x + 1) + K(x - 1)$ coincides with $\varphi(x)$ outside $[- 1, 1]$ and provides the m.e. to $[- 1, 1]$, consisting of the constant function 2/3 on $[- 1/2, 1/2]$ and linear pieces on $[- 1, - 1/2]$ and $[1/2, 1]$ (draw a picture).

<u>Exercise</u>. Compute the m.e. of

$$\varphi(x) = \begin{cases} (a - x)/(a - 1) & , \quad 1 \leqslant x \leqslant a \\ 0 & , \quad x \geqslant a \end{cases}$$

when φ is a) even, b) odd (consider separately the cases $1 < a \leqslant 2$, $2 < a \leqslant 3$, and $a > 3$).

<u>7.4.15 Use of Laplace transforms</u>. The solution of the difference equation for K which is the basis of the method developed above can be written in a compact form, useful for explicit computation, when φ is representable as a Laplace transform. Thus, suppose

(1) $$\varphi(x) = \int e^{-xs}\, d\sigma(s) \quad , \quad x \geqslant 1$$

where σ is a (not necessarily bounded) measure on $[0, \infty)$. Consider first the case of φ <u>even</u>. Then clearly

(2) $$K(x) = \int \frac{e^{-|x|s}}{e^{s} + e^{-s}}\, d\sigma(s)$$

satisfies $K(x + 1) + K(x - 1) = \varphi(x)$ for $x \geqslant 1$. If K is positive definite, then the usual computation gives

(3) $$\alpha(\varphi) = (4/\pi) \int \arctan e^{-s}\, d\sigma(s)$$

and

(4) $$F(x) = \int \frac{e^{xs} + e^{-xs}}{e^{2s} + 1}\, d\sigma(s)$$

is the m.e. on $[-1, 1]$.

In checking the positive definiteness of K, we can compute from (2) its Fourier transform k:

(5) $$k(t) = \int \frac{s}{e^{s} + e^{-s}} \cdot \frac{d\sigma(s)}{s^2 + t^2} \;.$$

This formal computation is certainly justified (by Fubini's theorem) if

(6) $$\int \frac{d|\sigma(s)|}{s(e^{s} + e^{-s})} < \infty \;,$$

hence if (6) holds, and $k(t) \geqslant 0$, (4) is a m.e. and (3) is valid.

We get a similar group of formulas for the <u>odd</u> case, starting from

(2′) $$K(x) = \int \frac{e^{-|x|s}}{e^{s} - e^{-s}}\, d\sigma(s)$$

which satisfies $K(x - 1) - K(x + 1) = \varphi(x)$ for $x \geqslant 1$. Because of the zero of $e^{s} - e^{-s}$ at $s = 0$, this formula is less likely to be useful than (2); however, the formally differentiated formula

(7) $$K''(x) = \int \frac{s^2}{e^{s} - e^{-s}} \cdot e^{-|x|s}\, d\sigma(s) \quad , \quad x \neq 0$$

does not suffer from this defect; in particular if the right side is non-negative, the essential hypothesis of 7.4.9 holds. In any case, if the function (2′) satisfies our usual sufficient conditions, then

(3′) $$\alpha(\varphi) = (2/\pi) \int \log \frac{e^{s} + 1}{e^{s} - 1}\, d\sigma(s)$$

and the m.e. on $[-1, 1]$ is given by

(4') $$F(x) = \int \frac{e^{xs} - e^{-xs}}{e^{2s} - 1}\, d\sigma(s) \quad .$$

__Exercises.__ a) Fill in the details of the above computations.

b) Derive anew the results of 7.4.13 with the aid of the above formulae.

__7.4.16 Further remarks on the preceding__

7.4.16.1 Let E_τ denote the set of (restrictions to the real axis of) entire functions of exponential type at most τ. It can be shown that $L^1 \cap E_\tau$ is a closed subspace of $L^1 = L^1(R)$. Consider now the problem, given $f \in L^1$, of finding a best approximation from the subspace $L^1 \cap E_\tau$. The n.a.s.c. that $g \in L^1 \cap E_\tau$ is a b.a. is that $f - g$ be orthogonal to $L^1 \cap E_\tau$. Now, a variant of the Paley-Wiener theorem (cf. BOAS) says that this subspace maps __via__ Fourier transformation onto $\{F \in A(\hat{R}) : F(x) = 0$ outside $[-\tau, \tau]\}$. Thus, g __is a best approximation to__ f __from__ $L^1 \cap E_\tau$ __if and only if__ $\hat{g}$ __is a m.e. of__ $\hat{f}$ __relative to__ $\hat{R} \setminus (-\tau, \tau)$. Therefore, __the previous results are equivalent to problems of best__ L^1 __approximation by entire functions of exponential type__.

__Exercise.__ Determine the best approximation from $L^1 \cap E_\tau$ for each of the following functions

a) $e^{-|t|}$ b) $(1 + t^2)^{-1}$ c) $(\sin at/t)^2$,

where $a > \tau$.

7.4.16.2 A remarkable property of the classes $(C)^\wedge$ and $(S)^\wedge$ can be read off from the formulae 7.4.11(1) and 7.4.12(1). Assuming sufficient regularity of φ, we obtain upon differentiating these equations

$$\lim_{x \to 1-0} F^{(2n)}(x) = \varphi^{(2n)}(1), \ n = 1, 2, \ldots \quad .$$

In other words, __if the m.e. is differentiated an even number of times (which is possible at all points__ $x \neq 1$ __or__ $- 1$) __the resulting function extends continuously across__ $x = \pm 1$, __and is again a m.e. relative to__ E. This leads to yet another interesting interpretation of the m.e., namely F __is that__ C^∞ __function on__ $[-1, 1]$

whose even order derivatives at the end points match those of the given function φ, so now this many-faceted problem is seen to relate to two-point interpolation problems and "Lidstone series" (BOAS p. 172). If we consider the case of φ (and hence F) odd, the even-order derivatives of F vanish at the origin, and the determining conditions for F are the two-point interpolation:

$$F^{(2n)}(0) = 0 \quad , \quad F^{(2n)}(1) = \varphi^{(2n)}(1) \quad , \quad n = 0, 1, \dots \, .$$

We don't know to what extent these relations have counterparts in other problems of m.e.

7.4.16.3 Generalizations. In the earlier discussion, it is not theoretically necessary to distinguish "even" and "odd" cases; the whole analysis can be carried out (as is done by Sz. Nagy) directly for the signum function sgn cos $(t + \delta)$ where δ is a real parameter. In fact, if $p(t) = \sum_{-n}^{n} a_\nu e^{i\nu t}$ is any real trigonometric polynomial such that sgn $p(t)$ has mean zero, we can carry out the previous analysis mutatis mutandis, and are led to the search for positive definite solutions K of the difference equation $\sum_{-n}^{n} a_\nu K(x - \nu) = \varphi(x)$, or more generally solutions K such that kp^2 is a non-negative function, where k is the distributional Fourier transform of K. In principle, this method ought to be applicable to m.e. problems for other sets E, too. The general idea is to seek some distribution α whose Fourier transform is a function, the signum of which has its spectrum in E, and then look for positive definite solutions of $\alpha * K = \varphi$.

7.5 Some applications of m.e.

7.5.1 Bohr's inequality and integrating kernels. Let f denote a trigonometric polynomial (here, with real frequencies λ_ν), and mean zero:

$$(1) \qquad f(t) = \sum_{1}^{n} a_\nu e^{i\lambda_\nu t} \quad ; \quad |\lambda_\nu| \geq c, \quad \nu = 1, \dots n \, .$$

Let F denote that (unique) primitive of f having mean zero: $F(t) = \sum a_\nu (i\lambda_\nu)^{-1} e^{i\lambda_\nu t}$. The Bohr inequality is then

$$(2) \qquad \|F\|_\infty \leq (\pi/2c) \, \|f\|_\infty \quad ,$$

$\| \|_\infty$ denoting as usual the sup norm on R. The Bohr inequality is dual to that of Bernstein: in the former, roughly speaking, f has no low frequencies and we get an upper bound for its primitive; in the latter, f has no high frequencies and we get an upper bound for its derivative. The reason for this situation is that the operation of passing to a primitive [derivative] is expressible as convolution with a kernel in $L^1(R)$ (which is just what is needed to get estimates) precisely when this operation is restricted to functions without low [high] frequencies. Specifically, we have $F = f * k$, and hence $\|F\|_\infty \leqslant \|k\|_1 \|f\|_\infty$, where k is any element of L^1 such that $\hat{k}(x) = (-ix)^{-1}$ for $|x| \geqslant c$. Therefore, to prove (2) it is enough to show that $\alpha(\varphi) = \pi/2c$, where $\varphi(x) = 1/x$ for $x \in E$, $E = \hat{R} \setminus (-c, c)$. By homogeneity it is enough to prove this for $c = 1$, and this we have done in 7.4.13 a). Similarly, we get (2) with $\| \|_\infty$ replaced by $\| \|_p$, $1 \leqslant p < \infty$. The analogous estimate for the r-fold primitive is obtained in the same way, with $\Lambda(r)c^{-r}$ on the right side when r is odd, and $\tilde{\Lambda}(r)\, c^{-r}$ when r is even.

Exercises. a) Show under the above hypotheses that the conjugate $\tilde{F}$ of F satisfies

$$\|\tilde{F}\|_p \leqslant (\tilde{\Lambda}(1)/c)\, \|f\|_p \, .$$

b) Show that Bohr's inequality for $p = \infty$ and its generalizations to higher derivatives and conjugates are all best possible.

Actually, there is no need to restrict the analysis to finite trigonometric sums. The essential point is that any $k \in L^1(R)$ such that $\hat{k}(x) = (-ix)^{-1}$ for $|x| \geqslant c$ is an integrating (convolution) kernel for functions with no spectrum in $(-c, c)$. This is not trivial, and deserves a proof.

7.5.1.1 Theorem. If $f \in L^\infty(R)$, $0 \notin S =$ Spectrum (f), $k \in L^1(R)$ and $\hat{k}(x) = (-ix)^{-1}$ on an open neighborhood of S, then $k * f$ is absolutely continuous, and its derivative equals f a.e. Moreover, if S is a set of spectral synthesis the conclusion holds even if $\hat{k}(x) = (-ix)^{-1}$ for $x \in S$.

This result is rather delicate, since k can't have an integrable derivative (otherwise $\hat{k}(x)$ would be $o(1/x)$ at infinity).

<u>Proof of Theorem.</u> Suppose first $f \in C^1$, and f' bounded, and $\hat{k}(x) = (ix)^{-1}$ on an open nbhd. of S. Let $g = f - k * f'$. We claim $g = 0$, and for this (by a fundamental theorem of distribution theory) it suffices to show that each $y \in R$ has a neighborhood N_y such that $g * h = 0$ whenever $h \in L^1 \cap C^2$ and $\hat{h}$ vanishes outside N_y.

Suppose first $y \notin S$. Then y has a nbhd. N_y which misses S, and hence also the spectrum of f', which is a subset of S. Therefore, if supp $\hat{h} \subset N_y$, $f * h$ and $f' * h$ vanish and hence $g * h = 0$. If $y \in S$, by hypothesis there is a nbhd. N_y on which $\hat{k}(x) = ix$, and hence if supp $\hat{h} \subset N_y$,

$$g * h = f * h - k * f' * h = f * h - k * f * h' = f * (h - k * h') = 0$$

because $h - k * h'$ has the Fourier transform $\hat{h}(x) - \hat{k}(x)(-ix)\hat{h}(x) = 0$, since $\hat{k}(x)(-ix) = 1$ on supp $\hat{h}$. Thus we have shown that $f = k * f'$.

For the general case, we use a standard "regularization" argument. Let $\{w_n\}$ be an approximate identity consisting of very nice functions, so that each $f_n = f * w_n$ is C^1 with a bounded derivative. Then by the above

$$f_n = k * f_n' = (k * f_n)'$$

$$\int_a^b f_n(t)dt = (k * f_n)(b) - (k * f_n)(a)$$

and, letting $n \longrightarrow \infty$ and using the fact that $f_n \longrightarrow f$ boundedly a.e. we deduce

$$\int_a^b f(t)dt = (k * f)(b) - (k * f)(a)$$

showing that $k * f$ is an integral of f, as was to be proved.

As to the final assertion of the theorem, suppose S is a set of spectral synthesis, and let the closed interval $[-\delta, \delta]$ be disjoint from S. Let g be any element of L^1 such that $\hat{g}(x) = (-ix)^{-1}$ for $|x| \geq \delta$. Then $\hat{k} - \hat{g}$ vanishes on S, and by the synthesis property there exists, for each n, $h_n \in L^1(R)$ such that $\hat{h}_n$ vanishes on an open nbhd. of S, and $\|k - g - h_n\|_1 < 1/n$. Thus, the sequence $k_n = g + h_n$ tends to k in L^1, and since $\hat{k}_n(x) = (-ix)^{-1}$ on an open nbhd. of S, $k_n * f$ is a primitive of f:

$$\int_a^b f(t)dt = (k_n * f)(b) - (k_n * f)(a) .$$

Finally, letting $n \to \infty$ and observing that $k_n * f$ converges uniformly to $k * f$, we are done. ◇

Exercise. Prove the above theorem by distribution theory.

As a consequence, if $g = \hat{G}$ denotes the m.e. of the function $\varphi(x) = 1/x$ from $R \setminus (-c, c)$, then $k = ig$ is an "integrating kernel" for all $f \in L^\infty$ whose spectrum misses $(-c, c)$. Hence

7.5.1.2 Corollary. If $f \in L^\infty$ and Spectrum (f) is disjoint from $(-c, c)$, f possesses a primitive function F, also with spectrum disjoint from $(-c, c)$, which satisfies $\|F\|_\infty \leq (\pi/2c)\, \|f\|_\infty$. Here $\pi/2c$ cannot be replaced by a smaller number, in fact equality holds for suitable f.

(Observe that the requirement $0 \notin$ Spectrum (F) uniquely determines a certain primitive.) Equality is attained if f is a constant times a translate of sgn cos ct.

Exercises. a) Sketch the graph of F for $f(t) = \operatorname{sgn} \cos t$.

b) State the appropriate generalizations to higher derivatives, L^p norms, and conjugate functions. Describe all the extremal functions f for which equality is attained.

c) Prove (what we have been taking for granted for a long time now!) that there exists $g \in L^1(R)$ such that $\hat{g}(x) = 1/x$, $|x| \geq 1$. Give two proofs, based (i) on the behaviour of the derivatives of $1/x$, (ii) on the homogeneity of $1/x$. Prove also the analogous results for $|x|^{-s}$ and $(\operatorname{sgn} x)|x|^{-s}$, $s > 0$.

The previous discussion of "integrating kernels" can be paraphrased for other operators. For instance, if $k \in L^1(R)$ and $\hat{k}(x) = ix$ on Spectrum (f), k is a "differentiating kernel" for f, $k * f = f'$. Such k can only exist, of course, if f has compact spectrum.

Exercise. Formulate and prove the analog of Cor. 7.5.1.2 for derivatives; also for the operator $f(t) \longrightarrow f(t+1) - f(t)$ and its inverse.

7.5.1.3 Digression on the philosophy of "multipliers". This is perhaps as good a point as any to remark that one of the most important techniques in modern analysis is the "multiplier technique", of which the above are simple illustrations (we shall encounter further examples elsewhere in these lectures). The general idea is this. Suppose we are given some distributions $\alpha, \beta_1, \dots \beta_r$ and we want an estimate, valid for f in some appropriate normed space H of functions on R^n, of $\alpha * f$ in terms of the corresponding norms of $\beta_1 * f, \dots, \beta_r * f$. Suppose for simplicity that $\hat{\alpha}$ and the $\hat{\beta}_i$ are (identified with) functions on $\hat{R}^n$ (certainly the case when α, β_i are derivations). Suppose also that we have some insight into the "multiplier space" M_H associated with H; by this we mean the set of functions ψ on $\hat{R}^n$ such that $\psi(x)\hat{f}(x)$ is the Fourier transform of some $g \in H$ whenever $f \in H$, and an inequality $\|g\| \leqslant A \|f\|$ holds for some $A = A_\psi$. The inf of the allowable A_ψ is the multiplier norm of ψ. The "multiplier technique" consists of looking for an identity

$$(1) \qquad \hat{\alpha}(x) = \sum_{i=1}^{r} \psi_i(x)\, \hat{\beta}_i(x) , \qquad x \in \hat{R}^n$$

with $\psi_i \in M_H$. This then allows the estimate

$$(2) \qquad \|\alpha * f\|_H \leqslant \sum_{i=1}^{r} \|\psi_i\| \, \|\beta_i * f\|_H ,$$

where $\|\psi_i\|$ denotes the multiplier norm of ψ_i. Of course, a representation of the form (1) is not unique when $r \geqslant 2$, or when $n = 1$ and $\hat{\beta}_1(x)$ vanishes on an open set, and one then tries to find, in some sense, a "minimal" solution, that is ψ_i satisfying (1) whose multiplier norms are in some sense as small as possible. The reader interested in details can find applications to Kolmogorov's inequality in DOMAR_2, KALLIONIEMI; to a priori estimates for differential operators in DE LEEUW & MIRKIL, BOMAN; to trigonometric approximation in SHAPIRO_3 (cf. also the next chapter). One of the reasons that Fourier-Stieltjes transforms turn up often in this work is that they are bounded multipliers for a vast array of spaces, including all $L^p(R^n)$ with

$p \geqslant 1$, in fact all "homogeneous Banach spaces" (cf. Chapter 9 below). ($L^p(R^n)$ with $1 < p < \infty$ admits, however, additional multipliers, cf. HÖRMANDER$_2$). An interesting possibility in the multiplier technique, which we have encountered in connection with differentiating and integrating kernels, is that (1) may not be satisfiable for all x, but only for x in some set E. Then we cannot expect an estimate (2) for all $f \in H$, but only for f whose spectrum lies in E. The search for ψ_i which are a "minimal solution" in a suitable sense, is then a generalization of the minimal extrapolation problem, to which it reduces for $r = 1$.

7.5.2 Theorem. (Favard, Ahiezer-Krein). Let $f \in C^{(r-1)}(R)$ where $r \geqslant 1$, be of period 2π with $f^{(r-1)} \in \mathrm{Lip}\,1$. Then, there exists for each n a trigonometric polynomial P_{n-1} of degree at most $n - 1$ such that

$$\|f - P_{n-1}\|_\infty \leqslant \lambda_r \|f^{(r)}\|_\infty n^{-r}$$

where $\lambda_r = \Lambda(r)$ for r odd, $\lambda_r = \tilde{\Lambda}(r)$ for r even (cf. 7.4.13).

Proof. Let s_{n-1} denote the partial sum, of order $n - 1$, of the Fourier series of f, and let $K_{r,n} = \hat{k}_{r,n}$ denote a minimal extrapolation of $\varphi(x) = (-ix)^{-r}$, $|x| \geqslant n$. Now $f - s_{n-1}$ has no spectrum in $(-n, n)$ and is the unique r-fold primitive having this property of $(f - s_{n-1})^{(r)}$. Hence, by the material of the previous section,

$$f - s_{n-1} = k_{r,n} * (f - s_{n-1})^{(r)} = k_{r,n} * f^{(r)} - Q_{n-1}$$

where $Q_{n-1} = k_{r,n} * s_{n-1}^{(r)} \in \mathcal{T}_{n-1}$ (trig. poly. of order at most $n - 1$). Hence $P_{n-1} = s_{n-1} - Q_{n-1} \in \mathcal{T}_{n-1}$ satisfies $\|f - P_{n-1}\|_\infty \leqslant \|k_{r,n}\|_1 \|f^{(r)}\|_\infty$ and the rest follows from 7.4.13. ◇

There is an analogous result for approximation to non-periodic functions by entire functions of exponential type. We formulate it in the "spectral" version:

Let $f \in C^{(r-1)}$, where $r \geqslant 1$, satisfy $f^{(r)} \in \mathrm{Lip}\,1$. Then, there exists for each real $c > 0$ a bounded continuous function F_c on R whose spectrum lies in $[-c,c]$ such that

$$\|f - F_c\|_\infty \leqslant \lambda_r \|f^{(r)}\|_\infty c^{-r}$$

<u>where λ_r is as in Theorem 7.5.2.</u>

(Of course, F_c must necessarily be (the restriction to R of) an entire function of exponential type c at most ("Paley-Wiener theorem").)

The proof is a good illustration of the multiplier technique. Let us try to find elements $h = h_{c,r}$ and $k = k_{c,r}$ of $L^1(R)$ such that the relation

(1) $$f = f * h + f^{(r)} * k$$

holds identically, and moreover

(2) $$\hat{h}(x) = 0 \quad , \quad |x| \geq c \; .$$

By the same kind of argument we used to prove Theorem 7.5.1.1, we see that (1) holds provided that

(3) $$1 = \hat{h}(x) + (-ix)^r \hat{k}(x) \quad , \quad \text{all } x \in \hat{R} \; .$$

Hence (1) and (2) both hold if

(4) $$\hat{k}(x) = (-ix)^{-r} \quad , \quad |x| \geq c \; .$$

Let us therefore choose $k = k_{c,r}$ to be <u>a solution of (4) having smallest norm</u>. It is easy to see that (3) then uniquely determines an element $h = h_{c,r}$ of $L^1(R)$ satisfying (2), hence $F_c = f * h_{c,r}$ has its spectrum in $[-c, c]$ and from (1),

$$\|f - F_c\|_\infty \leq \|f^{(r)} * k_{c,r}\|_\infty \leq \|f^{(r)}\|_\infty \, \|k_{c,r}\|_1 = \|f^{(r)}\|_\infty \cdot \lambda_r \, c^{-r} \; . \quad \diamond$$

<u>7.5.3 Approximating a function holomorphic in a strip</u>. The following example is taken from AHIEZER$_1$, p. 214. Let S_a denote the strip $\{|u| < a\}$ in the complex $w = t + iu$ plane, and denote by U_a the class of functions f holomorphic in S_a, real on the real axis, and satisfying

$$|\mathrm{Re}\, f(w)| \leq 1 \quad , \quad w \in S_a \; .$$

Such f have, a.e. on the boundary of S_a, non-tangential limiting values, hence

$$h(t) = \lim_{u \to a} \mathrm{Re}\, f(t + iu) = \lim_{u \to -a} \mathrm{Re}\, f(t - iu)$$

exists a.e., and $\|h\|_\infty \leq 1$. It is not hard to show (exercise!) that

$$(1) \qquad f(t) = (2a)^{-1} \int_{-\infty}^{\infty} h(s)(\cosh\ \pi(t-s)/2a)^{-1}\, ds$$

and that, conversely, every $h \in L^{\infty}$ with $\|h\|_{\infty} = 1$ determines an element of U_a by (1). In other words, the map $h \longrightarrow k * h$, where $k(t) = (2a \cosh(\pi t/2a))^{-1}$, takes the unit ball of L^{∞} one-to-one onto U_a. Now,

$$(2) \qquad \hat{k}(x) = (\cosh ax)^{-1}$$

(cf. OBERHETTINGER p. 34). Let now $G_c = \hat{g}_c$ denote a minimal extrapolation of $\hat{k}$, relative to $R \setminus (-c, c)$. Then

$$f_c = h * (k - g_c)$$

has its spectrum in $[-c, c]$, $\hat{k}(x) - \hat{g}_c(x) = 0$ for $|x| \geq c$. Moreover,

$$(3) \qquad \|f - f_c\|_{\infty} = \|h * g_c\|_{\infty} \leq \|h\|_{\infty} \|g_c\|_1$$

$$= (4/\pi)((\cosh ac)^{-1} - (3 \cosh 3ac)^{-1} + \ldots) < 4/\pi\, (\cosh ac)^{-1},$$

by 7.4.13 e). The conclusion is, therefore: <u>each</u> $f \in U_a$ <u>can be uniformly approximated on the whole real axis, by an entire function</u> f_c <u>of exponential type at most</u> c, <u>with an error given by (3), and in particular less than</u> $(4/\pi)(\cosh ac)^{-1}$. Of course, as is usual with this method, the constant is best possible. Qualitatively, the important thing to observe is that the error of the approximation is like const $\cdot\, (e^{-a})^c$, and the exponential decay with respect to the parameter c is the reflection, in approximation-theoretic terms, of the analyticity of f in a strip.

The above analysis, and that of the preceding paragraph, show that interesting subclasses of "nice" functions in L^{∞} are representable as the image of the unit ball under the map induced by a suitable convolution kernel. The approximation properties of that class <u>vis à vis</u> entire functions of exponential type are then known, if we can solve the m.e. problem for the relevant kernel, with respect to the sets $R \setminus (-c, c)$.

<u>Exercises</u>. a) Prove that the map (1) takes the unit ball of L^{∞} one-to-one onto U_a.

b) Extend the above analysis to handle functions f that are not necessarily real on the real axis.

c) Show that (3) is best possible, and determine all f for which equality can hold.

d) For a given continuous function f of period 2π, define $E_n(f) = \text{dist}(f, \mathcal{T}_{n-1})$. Show that the number $\beta = \overline{\lim}\, E_n(f)^{1/n}$ equals e^{-b}, where b is the supremum of the set of numbers a such that f is extendible as a bounded analytic function to the strip $\{|\text{Im } w| < a\}$.

e) Show that $f \in C(-1, 1)$ satisfies $\overline{\lim\limits_{n \to \infty}}\, \text{dist}(f, \mathcal{P}_{n-1})^{1/n} = \beta$ if and only if f is extendible as an analytic function to a certain ellipse E_β with foci at ± 1. What are the semi-axes of this ellipse?

f) Prove that for each $k \in L^1(R)$ the map $f \longrightarrow k * f$ takes the unit ball of L^∞ onto a compact subset of L^∞.

__7.6 Functions with a spectral gap__. As we have seen, the solution of minimal extrapolation problems is closely connected with the study of unimodular functions whose spectrum lies in a given set. In this section we wish to discuss just one case of this problem. Actually, it is more natural to study the more general problem of __arbitrary__ measurable functions bounded by 1, with spectrum in a given set, although the subclass which has modulus 1 everywhere, or at least on a set of positive measure, is of special interest in connection with the m.e. problem. (Of course, the study of __unbounded__ functions, and even of tempered distributions, with spectrum in a given set is a matter of great interest and importance; but since not a great deal is known about this as yet, and the subject leads far afield from approximation theory, we shall not discuss it further here). In this section we shall study only the case of __bounded functions whose spectra lie in__ $E_c = R \setminus (-c, c)$. So far as we know, the only detailed study of this class has been made by Benjamin Logan, Jr., in a notable 1965 doctoral dissertation (in electrical engineering!) Logan calls such functions "high pass" functions, a terminology based upon their significance in certain questions of electrical filter theory. More generally, a function in $L^\infty(R)$ (Logan actually studies a wider class) has, in Logan's terminology, a __spectral gap__, if the support of its spectrum is not all of $\hat{R}$. Upon multiplication by a suitable exponential e^{iat}, the "gap" can be made to include the origin. Qualitatively, functions with a spectral gap encompassing 0 are distinguished by <u>oscillatory behaviour</u>, and the larger the gap the more rapid the oscillations, a prototypical example being cos ct.

By way of orientation, we mention first these results of Logan:

(i) A non-null function f with a spectral gap cannot vanish on a half-line; indeed, if even $\int_0^\infty |f(t)|\, e^{\epsilon t}\, dt < \infty$ for some $\epsilon > 0$, and f has a spectral gap, then $f = 0$. (A function with a gap can, however, vanish non-trivially on arbitrarily long intervals.)

(ii) A non-null real-valued function f with a spectral gap about 0 cannot be of constant sign on a half-line.

We shall not prove these facts here; both can be deduced quite easily with the aid of the Fourier-Carleman transform. Instead we turn to other results; nearly everything in this section is contained in LOGAN, although we do not always give his proofs.

<u>7.6.1 Definition</u>. $H_c = \{f \in L^\infty : \text{Spectrum } (f) \cap (-c, c) = \emptyset\}$. It is easy to see that, equivalently,

$$H_c = \left\{f \in L^\infty : \int f(t)\, g(-t)dt = 0 \quad \text{whenever} \quad g \in L^1,\ \text{supp}\, \hat{g} \subset [-c, c]\right\}$$

Here we can also replace $\int f(t)\, g(-t)dt = 0$ by $f * g = 0$, because of the translation-invariance of H_c. Finally, we leave it as an exercise to show that H_c is precisely the set of bounded solutions of the integral equation

$$\int_{-\infty}^{\infty} f(t-u)(\sin cu/u)^2\, du = 0 \quad .$$

<u>7.6.2 Theorem</u>. <u>If</u> $f \in H_c$, <u>and</u> $-\infty < a < b < \infty$, <u>then</u>

$$\left| \int_a^b f(t)\, dt \right| \leq (\pi/c)\, \|f\|_\infty$$

and the constant π is best possible.

Proof. By Bohr's inequality, f possesses a primitive F bounded by $(\pi/2c)\ \|f\|_\infty$. Hence

$$\left|\int_a^b f(t)dt\right| = |F(b) - F(a)| \leq (\pi/c)\ \|f\|_\infty \ .$$

The function $f(t) = \operatorname{sgn} \sin ct$ shows the π is sharp. ◇

7.6.2.1 Corollary. If $f \in H_c$ and $\|f\|_\infty = 1$, we cannot have $f(t) = 1$ throughout an interval of length greater than π/c.

As a consequence, we see that if $G = \hat{g}$ is a m.e. relative to $E_c = R \setminus (-c,c)$, g cannot have constant sign throughout an interval of length greater than π/c.

We come now to an interesting result of Logan. The proof is a nice application of results developed earlier on m.e.

7.6.3 Theorem. Necessary and sufficient that $f \in L^\infty$ belong to H_c is

$$(1) \qquad |F(t,y)| \leq A\ e^{-cy}$$

with A independent of y, where

$$(2) \qquad F(t,y) = (y/\pi) \int_{-\infty}^{\infty} ((t-u)^2 + y^2)^{-1} f(u)\ du$$

is the Poisson integral of f.

Before turning to the proof, we would like to point out an interesting consequence of the theorem.

7.6.3.1 Corollary. If a function G is harmonic and bounded in the upper half-plane, and for every $c > 0$, $G(x + iy) = O(e^{-cy})$ uniformly in x, as $y \longrightarrow \infty$, then $G = 0$.

Indeed, G is the Poisson integral of a certain $g \in L^\infty(R)$, and Theorem 7.6.3 implies g has no spectrum in $(-c, c)$. Since c can be taken arbitrarily large, Spectrum (g) is empty, hence $g = 0$, $G = 0$. ◇

7.6.3.1 is a well-known "Phragmén-Lindelöf theorem", and easily proved by standard methods of complex analysis; however, we believe the present method of proof to be of some interest. Moreover, it can be used with only slight modifications to prove the corresponding theorem for bounded harmonic functions in a half-space of R^n. One can also establish by other methods (e.g. a Tauberian remainder theorem of GANELIUS) the following stronger proposition: <u>if</u> G <u>is harmonic and bounded in the upper half-plane, and for every</u> $c > 0$, $G(iy) = O(e^{-cy})$ <u>as</u> $y \longrightarrow \infty$, <u>then</u> $G(iy) = 0$ <u>for all</u> $y > 0$ (or, what is the same thing, G <u>is the Poisson integral of a bounded odd function on</u> R).

<u>Proof of Theorem 7.6.3.</u>

a) <u>Necessity</u>. Suppose $f \in H_c$. Then, writing

$$P_y(t) = (y/\pi)(t^2 + y^2)^{-1}$$

we have

$$F(t, y) = \int_{-\infty}^{\infty} f(t - u)\ P_y(u)\ du = \int_{-\infty}^{\infty} f(t - u)\ k_y(u)\ du$$

for any $k_y \in L^1$ such that $\hat{k}_y(x) = \hat{P}_y(x) = e^{-y|x|}$ for $|x| \geqslant c$. In particular, we may choose $\hat{k}_y$ to be a m.e. of $e^{-y|x|}$ relative to $E_c = \{|x| \geqslant c\}$, which gives, by 7.4.13 d)

(3) $$|F(t, y)| \leqslant \|f\|_\infty\ \|k_y\|_1 = \|f\|_\infty \cdot (4/\pi) \text{ arc tan } e^{-cy}$$

and (1) is established.

b) <u>Sufficiency</u>. We now suppose that (1) holds, in other words

(4) $$\|f * P_y\|_\infty \leqslant A\ e^{-cy}$$

and wish to show $f \in H_c$. For this it is sufficient to show that for each $b < c$, and each $g \in L^1$ with supp $\hat{g} \subset [-b, b]$, we have $f * g = 0$. Now, the convolution equation $P_y * h = g$ is solvable for $h = h_y \in L^1$, indeed the unique solution is that h which satisfies $\hat{h}(x) = \hat{g}(x)\ e^{y|x|} = \hat{g}(x)\ K_y(x)$, where now $K_y = \hat{k}_y$ denotes any function in $A(\hat{R})$ which equals $e^{y|x|}$ for $|x| \leqslant b$. We shall choose for K_y a m.e. of $e^{y|x|}$ relative to $[-b, b]$. It is easy to check (cf. 7.3.10) that when the function $\varphi(x) = e^{y|x|}$ is extended from $[-b, b]$ to the whole line so as to have period 2b, the periodic function $\psi(x) = \varphi(x + b)$ has non-negative Fourier coefficients (this is a consequence of the fact that ψ is even, and decreasing and convex on $[0, b]$). Thus ψ is positive definite, and the 2b-periodically extended φ is a m.e. (which is then the K_y referred to above), with norm e^{by}. Therefore

$$\|h\|_1 \leqslant \|g\|_1 \, \|k_y\|_1 = \|g\|_1 \, e^{by} ,$$

and so

$$\|f * g\|_\infty = \|f * P_y * h\|_\infty \leqslant \|f * P_y\|_\infty \, \|h\|_1 \leqslant A \, e^{-cy} \, e^{by} .$$

Since this holds for all y, we conclude, letting $y \longrightarrow \infty$, that $f * g = 0$, and the theorem is proved. ◇

From the above argument we can extract the following theorem of Phragmén-Lindelöf type, which we have not seen in the literature.

7.6.4 Theorem. Let F be a function harmonic in $\{y > 0\}$, and suppose

(i) $$|F(x, y)| \leqslant M \quad , \quad y > 0 ,$$

(ii) for every $\epsilon > 0$, there is a constant A_ϵ independent of x such that

(1) $$|F(x, y)| \leqslant A_\epsilon \, e^{-(c-\epsilon)y} .$$

Then

(2) $$|F(x, y)| \leqslant (4/\pi) \, M \arctan e^{-cy} < (4/\pi) \, M \, e^{-cy}$$

for all (x, y) in the upper half-plane.

The proof is immediate. Letting $f(x) = \lim_{y \to 0} F(x, y)$ a.e., so that F is the Poisson integral of f, (1) implies, in view of the preceding theorem, that for every $\epsilon > 0$, Spectrum (f) is disjoint from $(-c + \epsilon, c - \epsilon)$, and hence $f \in H_c$. The desired conclusion now follows from 7.6.3 (3). ◇

Under the hypotheses of Theorem 7.6.4 we can also get a bound for the harmonic function G which is conjugate to F, namely

(3) $$|G(x, y)| \leqslant (2/\pi) \log \frac{e^{cy} + 1}{e^{cy} - 1}$$

The proof is very similar to that of the preceding result, except that now we use the m.e. of $(\operatorname{sgn} x) \, e^{-y|x|}$, and we skip the details.

7.6.5 Real unimodular functions in H_c. Using the estimates 7.6.4 (2),(3), which hold whenever $f \in H_c$, LOGAN found a remarkable "parametric representation" of $\pm$ 1-valued functions in H_c. To discuss this result and some ramifications thereof, it will be convenient to first make some definitions.

7.6.5.1 Definitions. (N) denotes the class of functions $f(z)$, $z = x + iy$ holomorphic in $\{y > 0\}$ and satisfying there $\operatorname{Im} f(z) \geqslant 0$. (N_b) denotes the set of measurable complex-valued functions ψ on R which are boundary values of functions of class (N), thus $\psi(x) = \lim_{y \to 0} f(x + iy)$ (it is known that for $f \in (N)$ this limit exists a.e.). Finally, (N_b^r) denotes the set of real-valued functions in (N_b).

The N here stands for R. Nevanlinna, who introduced this class in connection with the study of interpolation and moment problems, and found for (N) the following "parametric representation" (analogous to, and in fact easily deduced from, the Riesz-Herglotz representation for the analogous class in the unit disc, cf. AHIEZER_2 p. 92).

7.6.5.2 The necessary and sufficient condition that a function f holomorphic for $\{y > 0\}$ belong to (N) is that there exist constants $a \geqslant 0$, b real, and a positive bounded measure ρ on R such that

$$f(z) = az + b + \int [(1 + \xi z)/(\xi - z)]\, d\rho(\xi) \ . \tag{1}$$

Moreover, for $f \in (N)$, a, b and ρ in (1) are uniquely determined.

Since we see from (1) that

$$\operatorname{Im} f(x + iy) = ay + \int \frac{y}{(x - \xi)^2 + y^2} \cdot (1 + \xi^2)\, d\rho(\xi) \tag{2}$$

and the second term is the Poisson integral of the measure $\pi(1 + \xi^2)\, d\rho(\xi)$, it follows from the theory of the Poisson-Stieltjes integral that $\psi(x) = \lim_{y \to 0} f(x+iy)$ is real valued if and only if the measure $(1 + \xi^2)\, d\rho$ (or what comes to the same thing, $d\rho$) is purely singular with respect to Lebesgue measure. Hence, $\psi \in (N_b^r)$ if and only if

$$\psi(x) = ax + b + \lim_{y \to 0} \int \frac{(1 + \xi x)(\xi - x)}{(\xi - x)^2 + y^2}\, d\rho(\xi) \tag{3}$$

a.e., where ρ is a bounded positive measure on R which is purely singular with respect to Lebesgue measure. Formally we get from (3): $\psi \in (N_b^r)$ iff

(4) $$\psi(x) = ax + b + \int \frac{1 + x\xi}{\xi - x} d\rho(\xi)$$

where the integral must be understood as a singular integral. For x not in the support of ρ however, (4) is meaningful in the ordinary sense, and defines $\psi(x)$. Clearly ψ is infinitely differentiable, even analytically extendible, in the neighborhood of a point not in supp ρ. Observe that (2) implies the relations

(5) $$\operatorname{Im} f(x + iy) \geqslant ay \quad , \quad f \in (N)$$

where

(6) $$a = \lim_{y \to +\infty} y^{-1} \operatorname{Im} f(x + iy) \ ,$$

the limiting existing, uniformly on compact x-intervals, whenever $f \in (N)$.

7.6.5.3 Inner functions. An inner function g in $\{y > 0\}$ is a bounded analytic function satisfying $\lim_{y \to 0} |g(x + iy)| = 1$ a.e. It can be shown (see DUREN) that every inner function g in $\{y > 0\}$ is uniquely representable in the form

(1) $$g = Be^{if}$$

where B is a convergent (possibly empty) Blaschke product, and $f \in (N)$. By a Blaschke product we mean here a function

(2) $$B(z) = \prod_{n=1}^{\infty} \frac{1 - z/z_n}{1 - z/\bar{z}_n}$$

where $\operatorname{Im} z_n > 0$ and (condition for convergence)

(3) $$\sum |\operatorname{Im} z_n^{-1}| = \sum y_n / |z_n|^2 < \infty \quad , \quad z_n = x_n + iy_n \ .$$

If g is inner, the constant a in the formula 7.6.5.2 (1) (as applied to the f appearing in the canonical representation (1) of g) is called the logarithmic residue of g. It is not hard to show (cf. HEINS) that

(4) $$a = \lim_{y \to +\infty} - \log |g(x+iy)|/y$$

uniformly on compact x-intervals. (In particular, if g is a Blaschke product, the limit in (4) is zero.)

We now return to the study of high-pass functions.

7.6.5.4 Theorem. If $\psi \in (N_b^r)$ and its holomorphic extension f to the upper half-plane satisfies

(1) $$\lim_{y \to +\infty} y^{-1} \operatorname{Im} f(iy) \geqslant 1$$

then ψ is an "admissible change of variables" for each of the classes H_c, i.e. the composition $g \circ \psi$ belongs to H_c whenever g does.

Proof of Theorem 7.6.5.4. Suppose $g \in H_c$ and denote by G the Poisson integral of g. It is then easy to check that $J = G \circ f$ is the Poisson integral of $g \circ \psi$. Now, by 7.6.3, $|G(w)| \leqslant Ae^{-c\operatorname{Im} w}$ for $\operatorname{Im} w > 0$, hence

$$|J(z)| = |G(f(z))| \leqslant Ae^{-c\operatorname{Im} f(z)} \leqslant Ae^{-cy}$$

by 7.6.5.2 (5), where $z = x + iy$ and, again by 7.6.3, we conclude $g \circ \psi \in H_c$. ◇

A related "change of variables" result is

7.6.5.5 Theorem. Let g have period 2π and belong to H_c. Let ψ be a real-valued measurable function on R such that $e^{i\psi}$ is the boundary value of an inner function in the upper half-plane with logarithmic residue at least a. Then $g \circ \psi$ belongs to H_{ac}.

Proof. As in the preceding proof, suppose $g \in H_c$ and let G denote the Poisson integral of g. Let φ denote the inner function whose boundary values are $e^{i\psi}$. The function $\log \varphi(z)$ is a multiple-valued holomorphic function in the upper half-plane with logarithmic singularities at the zeroes of φ. Now, $\log \varphi$ is a single-valued holomorphic function plus the logarithm of its Blaschke factor, and from the form of the latter we deduce that at any two points of the Riemann surface of $\log \varphi$ which lie over the same point of the z-plane the values of φ differ by an integer multiple of $2\pi i$. Now, since g has period 2π, so has G, and consequently $G(-i \log \varphi(z))$ is indefinitely continuable as a harmonic function in $\{\operatorname{Im} z > 0\}$ as long as we avoid the zeroes of φ, and <u>single-valued</u>. This harmonic function being also <u>bounded</u>, its "apparent singularities" are removable, in other words there is a bounded harmonic function $J(z)$ in $\{\operatorname{Im} z > 0\}$ which agrees with $G(-i \log \varphi(z))$ on $\{z: \varphi(z) \neq 0\}$. Moreover, $J(z)$ is clearly the Poisson integral of $g(-i \log \varphi(x)) = g(\psi(x))$ (because g has period 2π, it doesn't matter what determination of $\log \varphi(u)$ we select). Now the proof concludes like that of the last theorem: since $g \in H_c$, we have $|G(w)| \leqslant Ae^{-c \operatorname{Im} w}$, hence

$$|J(z)| \leqslant Ae^{-c \operatorname{Im}(-i \log \varphi(z))} = Ae^{c \log|\varphi(z)|} = O(e^{-acy})$$

because φ has logarithmic residue at least a, and so by 7.6.3, $g \circ \psi \in H_{ac}$. ◇

A simple example of an inner function φ with logarithmic residue 1 is $\varphi(z) = e^{iz}(1 + iz)/(1 - iz)$. Here an associated ψ is $\psi(x) = x + 2 \arctan x$, and so we can conclude, taking for instance $g(t) = \operatorname{sgn} \cos t$, that the step function $\operatorname{sgn} \cos (x + 2 \arctan x)$ belongs to H_1. This would be rather troublesome to verify by direct computation. Observe that, having found a fairly simple step function in H_1, it is natural to look for new m.e. problems relative to $\hat{R} \setminus [-1, 1]$ which are explicitly solvable. Logan does just this, using the fact that

(1) $$g(x) = \operatorname{sgn} \cos (cx + 2 \arctan ax)$$

is in H_c to find explicitly

$$\sup \left| \int_{-T}^{T} h(t)dt \right| \quad , \quad h \in H_c \quad , \quad \|h\|_\infty = 1 .$$

Indeed, for $T \geqslant \pi/2c$ he shows that the extremal h is the function g given by (1), where a is so adjusted that g changes sign at $x = \pm T$. However, we do not wish to enter here into the details of this determination, but rather go on to the following general theorem due to Logan (in a slightly different formulation).

<u>7.6.5.6 Theorem (LOGAN)</u>. <u>Let</u> h <u>be a measurable function on</u> R <u>such that</u> $h(x)^2 = 1$ <u>a.e.</u> <u>Then</u> $h \in H_c$ <u>if and only if</u>

(1) $$h(x) = \operatorname{sgn} \cos \psi(x)$$

<u>where</u> $e^{i\psi(x)}$ <u>is the boundary value of an inner function in the upper half-plane with</u>

<u>logarithmic residue at least</u> c.

<u>Proof</u>. That (1) implies $h \in H_c$ is a consequence of the preceding theorem. Let us therefore turn to the converse assertion. We suppose $h \in H_c$, and let U denote the Poisson integral of h, and V its harmonic conjugate, so normalized that the function $F = U + iV$, holomorphic for $\{y > 0\}$ satisfies

(2) $$F(x + iy) \leqslant Ae^{-cy} \quad , \quad y \geqslant 1$$

which we know as possible by 7.6.3 and 7.6.4 (3). Observe that, by the maximum principle

(3) $$|U(x + iy)| \leqslant 1 \quad , \quad y > 0 \ .$$

Consider now the function $J = e^{i(\pi/2)F}$, i.e.

(4) $$J(z) = e^{-(\pi/2)V(z)} \cos (\pi/2)U(z) + ie^{-(\pi/2)V(z)} \sin (\pi/2)U(z) \ .$$

Then J is holomorphic for $y > 0$ and, because of (3),

(5) $$\operatorname{Re} J(x + iy) > 0 \quad \text{for} \quad y > 0 \ .$$

Moreover, the boundary values $J(x)$ satisfy, as we see from (4) (recall that $U(x) = h(x) = \pm 1$ a.e.!)

(6) $$J(x) = ie^{-(\pi/2)\tilde{h}(x)} \sin (\pi/2)h(x) \quad \text{a.e.}$$

where $\tilde{h}$ is the conjugate function (Hilbert transform) of h. From (6) we get

(7) $$\operatorname{sgn} J(x) = i\, h(x) \quad \text{a.e.}$$

Consider now the function

(8) $$G(z) = \frac{1 - J(z)}{1 + J(z)}$$

which is holomorphic for $\{y > 0\}$ and has modulus bounded by one. We see from (7) that G is inner. From (7), (8) we get

(9) $$i\, h(x) = \operatorname{sgn} \frac{1 - G(x)}{1 + G(x)} \quad \text{a.e.}$$

Now, because of (2), $J(x + iy) = 1 + O(e^{-cy})$ as $y \longrightarrow \infty$, hence from (8), $G(x + iy) = O(e^{-cy})$ showing that G has logarithmic residue at least c. Finally, observing that for any complex number ω of modulus one

$$\operatorname{sgn}\,[(1-\omega)/(1+\omega)] = -i\,\operatorname{sgn}\,[\operatorname{Im}\omega]$$

we obtain from (9), if ψ_1 is chosen to be any real-valued measurable function on R such that $e^{i\psi_1(x)} = G(x)$ a.e.,

$$h(x) = -\operatorname{sgn}\operatorname{Im} G(x) = -\operatorname{sgn}\sin\psi_1(x) = \operatorname{sgn}\cos\psi(x)$$

where $\psi(x) = \psi_1(x) + \pi/2$ satisfies the requirements of the theorem. ◇

Remark. We can of course replace cos in (1) by any translate thereof, in particular, by sin.

This theorem allows us to obtain an explicit representation for all ± 1-valued step functions in H_c (by a step function we mean a piecewise constant function whose points of discontinuity have no finite limit point). For simplicity, we make the inessential restriction $c = 1$.

7.6.5.7 Theorem. The necessary and sufficient condition that a step function taking only the values 1 and - 1 belong to H_1 is that the set of points where it changes sign be the image of the set $\{n\pi\}$, $n = 0, \pm 1, \ldots$ under a homeomorphism $x \longrightarrow \psi(x)$ of R where

(1) $$\psi'(x) = a + 2\sum_{n=1}^{\infty} \frac{y_n}{(x-x_n)^2 + y_n^2} \quad ,$$

(2) $$a \geqslant 1$$

(3) $$\sum_{n=1}^{\infty} |\operatorname{Im}(1/z_n)| = \sum_{n=1}^{\infty} y_n/(x_n^2 + y_n^2) < \infty \; .$$

Proof. (i) Necessity. If h is such a step function, it is representable, by the preceding theorem, as

(4) $$h(x) = \operatorname{sgn}\sin\psi(x)$$

where $e^{i\psi}$ is the boundary value of an inner function φ with logarithmic residue at least one. Now, because h is a step function, (4) together with simple observations about the discontinuities of boundary values of inner functions permit us to conclude that $\varphi(z) = \omega e^{iaz} B(z)$, where ω is a constant, $a \geqslant 0$ and B is a Blaschke product whose zeroes $\{z_k\}$ do not cluster at any point of R. Moreover, $a \geqslant 1$. Now, since $\varphi(x)$ is continuous and has modulus one there exists a continuous real-valued function x satisfying $e^{i\psi(x)} = \varphi(x)$, and it is moreover clear that ψ is differentiable and

(5) $$i\psi'(x) = ia + (d/dx) \log B(x) \quad .$$

Substituting for B(x) from 7.6.5.3 (2) and carrying out a small computation gives

$$\psi'(x) = a + 2 \sum (y_n / |z_n - x|^2) \quad , \quad z_n = x_n + iy_n$$

and $x \longrightarrow \psi(x)$ is therefore a homeomorphism of R with the properties asserted.

(ii) The sufficiency follows by observing that the above argument is reversible. If ψ is defined (up to an additive constant) by (1), where the $z_n = x_n + iy_n$ satisfy (3) and $a \geqslant 1$, then the Blaschke product B formed with the z_n converges and (5) holds, implying in turn $e^{i\psi(x)} = \omega\, e^{iax} B(x)$, which by the preceding theorem implies that sgn sin $\psi(x) \in H_1$. This is equivalent to the assertion that was to be proved. ◇

Thus we have now pushed the "in depth" study of the m.e. problem for $\hat{R} \setminus [-1, 1]$ to the point where we have an explicit representation for the patterns of sign changes of all extremals $f \in L^1(R)$ which do not vanish on any set of positive measure and which do not change sign infinitely often in the neighborhood of any finite point. One might try to push further and deal with the case of extremals that are allowed to vanish on an interval, or what comes to the same thing, $\pm$ 1-valued step functions defined on a proper subset of R, which can be extrapolated to all of R so that the extrapolated function belongs to H_1 and has sup norm 1. We leave this, as well as all kinds of possible generalisations to other sets, higher dimensions, applications to specific extremal problems, etc., for future investigations. We wish, however, to close this chapter by a few remarks which clarify further the meaning of "spectrum", and in particular shed further light on high-pass functions.

7.6.6 Useful characterizations of the spectrum. In localizing the spectrum of a function f which does not grow too fast at infinity ("tempered" function, in L. Schwartz' sense; in this section we are in any case concerned only with bounded functions) there are two criteria which are often useful. The first interprets the spectrum in terms of analytic continuation (or, in a variant form, harmonic continuation) via the so-called Fourier-Carleman transform (cf. KATZNELSON, p. 179; KAHANE & SALEM p. 161 for the discrete analogue). The second is based on trying to interpret the spectrum in terms of "where the Fourier transform vanishes", which is the most natural way if our function is integrable (so that its F.T. is a continuous function), and in the general case writing the Fourier formula with a "summability kernel" that ensures convergence. The general principles governing these procedures are summed up in the following two propositions (neither of them trivial) which we state here without proof. For the first, cf. KATZNELSON; the second is similar to a lemma stated by HERZ.

A. For $f \in L^\infty(R)$, the complement of the spectrum of f is that open (possibly empty) subset of $\hat{R}$ across which the two analytic functions

$$F_+(z) = -\int_{-\infty}^{0} f(t)\, e^{-itz}\, dt \quad , \quad \operatorname{Im} z > 0$$

$$F_-(z) = \int_{0}^{\infty} f(t)\, e^{-itz}\, dt \quad , \quad \operatorname{Im} z < 0$$

are analytically continuable into one another.

B. (i) If $f \in L^\infty(R)$, and $k \in L^1(R)$ is bounded, continuous at 0, and satisfies $k(0) = 1$, and

$$(1) \qquad \lim_{\epsilon \to 0} \int_{-\infty}^{\infty} k(\epsilon t)\, f(t)\, e^{-ixt}\, dt = 0$$

uniformly for x in some neighborhood of x_0, then x_0 is not in the spectrum of f.

(ii) If $f \in L^\infty(R)$, and $k \in L^1(R)$ is absolutely continuous with an integrable derivative, then (1) holds uniformly for x in each compact set disjoint from the spectrum of f.

7.6.6.1 As an application of A, let f be a step function taking only the values ± 1 ("square wave", in engineering terminology). As we already know, f cannot take the same value on an interval of length greater than π/c, and consequently has infinitely many sign changes both as $t \longrightarrow -\infty$ and as $t \longrightarrow \infty$. By a translation we can arrange that $f(t) = 1$ in a neighborhood of 0. Let

$$-\infty < \dots -s_2 < -s_1 < 0 < t_1 < t_2 < \dots +\infty$$

denote the points where the sign changes occur. Then, according to A, $f \in H_c$ if and only if

(1) $$F_+(z) = (1 - 2e^{is_1 z} + 2e^{is_2 z} - \dots)/iz$$

and

(2) $$F_-(z) = (1 - 2e^{-it_1 z} + 2e^{-it_2 z} + \dots)/iz$$

are analytic continuations of one another across $(-c, c)$. From this we can infer that for such a step function to belong to any H_c with $c > 0$ is "rather unlikely", insofar as functions defined by Dirichlet series (as are (1) and (2)) are "rather unlikely" to be continuable analytically across any points of their abscissae of convergence. To take a more concrete case, if the s_n, say, are integer multiples of some fixed number δ, then $izF_+(z)$ is a power series in $e^{i\delta z}$ with integer coefficients and hence by the Pólya-Carlson theorem (cf. BIEBERBACH) nowhere continuable across the real axis except when the differences $s_{n+1} - s_n$ are ultimately periodic or, what is the same thing, izF_+ reduces to a rational function of $e^{i\delta z}$.

By application of B, with concrete choices of k (an interesting choice is $k(t) = (1 - |t|)^+$) we can get any number of further n.a.s.c. for high-pass functions.

7.6.7 Exercises. a) Use 7.6.5.2 to obtain a limitation on the rapidity of oscillations of a square wave in H_c.

b) (Logan) Let $f \in L^1(R)$ vanish outside the interval $[0, \pi/c]$. Prove that f is orthogonal to that subspace L^1_c of $L^1(R)$ consisting of functions whose Fourier transforms vanish for $|x| \geqslant c$. (Hint: let $\psi(t) = \operatorname{sgn} \overline{f(t)}$ where $f(t) \neq 0$, $\psi(t) = 0$ elsewhere on $[0, \pi/c]$. Now extend ψ to $[-\pi/c, 0]$ by oddness, and thereafter by periodicity to all of R. The thus extended function $\tilde{\psi}$ is bounded by one, belongs to H_c and satisfies $\psi(t)f(t) = |f(t)|$ a.e. Now apply 7.3.7.)

c) Prove the following reformulation of the assertion in b): if $F \in A(\hat{R})$ is (the restriction to the real axis of) an entire function of exponential type not exceeding $\pi/2c$, then F is a m.e. relative to $\hat{R} \setminus (-c, c)$.

d) (Logan) Prove that the conclusion in b) holds if $f \in L^1(R)$ and vanishes outside a set of measure $\pi/2c$ (Note: This is not a misprint; it is unknown whether one can here write π/c. Hint: Establish, for $g \in L^1_c$, the inequality $\|g\|_\infty \leq (c/\pi)\, \|g\|_1$.)

e) Find a sequence of numbers $0 < r_1 < r_2 < \dots$ such that the function ψ on R^2 equal to 1 for $|t| < r_1$, -1 for $r_1 \leq |t| < r_2$, and so on alternately, has a spectral gap about 0, i.e. supp $\hat{\psi}$ does not contain 0.

f) Use the criteria of 7.6.6 to determine the spectra of the functions cos t, sgn cos t, sin t^2, sgn sin t^2, sgn t.

Chapter 8. General Aspects of "Degree of Approximation"

8.1 "Best" vs "good" approximation; generalities. In previous chapters, we have considered the problem of best approximating a given element of a normed linear space X by an element from some subset S of X (in most cases, a subspace). We worked in fairly special spaces, such as $C(Q)$, L^p spaces, Hilbert spaces and $A(\hat{R})$. Now, as we have already intimated, a reasonably useful theory of "best approximation" can be carried out in the full generality of normed linear spaces, at least when S is a closed subspace, or more generally a closed convex set. In order, however, to get on into the other hemisphere of approximation theory, the theory of "good approximation", we shall not pursue this general line, but refer the reader to the exhaustive treatise of SINGER_1. In leaving the subject of "best approximation", we wish however to emphasize that we have only scratched the surface, and urge the interested reader to pursue for himself questions such as best approximation by non-convex families S (such as rational functions of fixed degree), and algorithms for computing best approximations, in addition to the above-mentioned generalizations to normed spaces. Among others, the books of CHENEY, MEINARDUS, RIVLIN, and RICE should prove helpful, the last-named especially for its coverage of non-linear approximation.

From one point of view, the problem of "good approximation" can be viewed in the same spirit as that of "best approximation" except that now we are trying to approximate not to a single element but to some class of elements. Specifically, if K, S are two subsets of the normed linear space X, let us consider the deviation of K from S, defined as

$$\text{dev}\,(K; S) = \sup_{x \in K} \text{dist}\,(x, S)\ .$$

When K reduces to a single element x_0, we are thus asking about the distance from x_0 to S, and the associated questions of existence, uniqueness, and characterization of closest elements. In the general case, we can inquire into the existence of an element of K whose distance from S equals dev (K; S), i.e. one which is "hardest to approximate" by elements of S, among all the elements of K. For example, if X is that subspace of C(R) consisting of 2π-periodic functions, $S = \mathcal{T}_{n-1}$ and

$K = \left\{f \in X: |f(t_2) - f(t_1)| \leq |t_2 - t_1|\right\}$ (the normalized Lip 1 class) we are asking how well a (normalized) Lip 1 function is approximable by trigonometric polynomials of degree n - 1. From the preceding chapter, we know the answer: always with an error of at most $\pi/2n$, and this bound is unimprovable, the "hardest to approximate" element being the "saw-tooth" function of period $2\pi/n$ whose graph consists of line segments of slope alternately ± 1. One line of investigation seeks to solve the analogous problem for various situations X, S, K. Another, somewhat more abstract twist can be given to the problem by keeping K fixed and varying S within a certain family (S). One thus asks about the number inf dev (K; S) as S ranges over (S). An important case of this is the Kolmogorov problem of "widths", whereby S ranges over all vector subspaces of X of fixed (finite) dimension. Thus, in the above example, it is known that no vector subspace of X of dimension 2n - 1 has a smaller deviation from K than $\pi/2n$, the deviation attained by the subspace $\mathcal{T}_{n-1}$. This may be interpreted as saying that the trigonometric polynomials of degree n - 1 afford the best approximation to Lip 1 functions that is possible for any <u>linear</u> family (i.e. linear vector space) with the same number of free parameters (= dimension = 2n - 1). This "geometrization" of the theory of approximation leads to characteristic perspectives, problems, and theorems, cf. TIHOMIROV$_1$, LORENTZ$_1$, SINGER$_1$.

Ultimately, however, if one is after <u>explicit</u> results, one must ask other questions too, insofar as experience shows that when the problems of approximation theory are formulated as <u>extremal problems</u> of one and another sort, these problems will be explicitly solvable, and yield numerical estimates, only in a relatively small number of cases. Although these cases are among the gems of mathematical analysis, we shall hereafter in these lectures ask questions in a different spirit, the spirit of seeking orders of magnitude, estimations, inequalities, and allowing "constants" into the analysis whose values it may be impossible for us to ascertain.

In general, we get led into asymptotic problems when we consider either a family S_λ of subsets of X, or a family A_λ of "approximation-generating operators" from X into itself, depending on, say, a real parameter λ which tends to $+\infty$. In the first situation, typically, the S_λ are nested, i.e. $S_\lambda \subset S_{\lambda'}$ when $\lambda < \lambda'$ and $\bigcup S_\lambda$ is dense in X. We then try to estimate, for given $x \in X$, the rate of decrease to zero of dist (x, S_λ) as $\lambda \to \infty$, or more generally, for given $K \subset X$, the behaviour of dev $(K; S_\lambda)$ as $\lambda \to \infty$. In the second (operator) situation, typically, the A_λ are linear operators (an "approximate identity" in some sense), and we study the rate of decrease to zero of $\|x - A_\lambda x\|$ as $\lambda \to \infty$, or more generally the behaviour of $\sup_{x \in K} \|x - A_\lambda x\|$ as $\lambda \to \infty$, for various classes K.

Typical questions governed by the first situation might be: What is the order of magnitude of the distance, in C(I), from the function $|x|$ to $\mathcal{P}_n$? How closely are convex functions on I approximable by elements of $\mathcal{P}_n$? What structural properties can we expect of a function that is approximable to the order $O(n^{-2})$ by polynomials of degree n? Typical questions governed by the second situation might be: How rapidly (if at all) do the Lagrange interpolating polynomials of the function $|x|$ on I, corresponding to equidistant points of subdivision, converge? How rapidly do the Fejér means of the Fourier series of Lip α functions converge? How do the Bernstein polynomials of an analytic function behave? What structural properties can we expect of a function $f \in L^p(R)$ that is approximable to the order λ^{-2} by its Weierstrass mean $\pi^{-1/2} \int_{-\infty}^{\infty} f(t - \lambda^{-1}u)\, e^{-u^2} du$?

Obviously, there is an almost limitless reservoir of questions, even more or less natural ones, that one is led into. Whole books can be (in fact, have been) written about individual aspects of these questions: interpolation, orthogonal polynomials, Fourier series, Taylor series, Bernstein polynomials, convolution integrals, spline approximation, approximation of analytic functions, etc. etc., covering topics which because of their practical importance or theoretical interest have inspired detailed investigations. In the remainder of these lectures, we shall, to achieve a certain unity of treatment, impose severe limitations on the scope of the investigation. For problems governed by the first situation (above) we shall deal almost exclusively with the case where X is a suitable Banach space of functions on R^n, and S_λ consists of trigonometric polynomials of degree not exceeding λ. (The methods used handle equally well the somewhat more general families S_λ of (non-periodic) entire functions of exponential type λ which are bounded on the real axis and, by change of variables from the periodic case, we can also deal with <u>algebraic</u> polynomials of fixed degree.) As for the second situation (operators) we shall study almost exclusively approximation operators A_λ of the special form $f \to \int f(t - \lambda^{-1} u)\, k(u)\, du$ where $k \in L^1(R^n)$ and $\int k\, dt = 1$. What is nice about studying this special kind of operator is that quite precise information can be obtained, and there is a unity of method, based upon the "multiplier technique" discussed in the preceding chapter. It turns out that much of the detailed behaviour of such an approximation operator is reflected in the behaviour of the Fourier transform of the "kernel" k near the origin, and powerful techniques from Fourier analysis can be used to advantage. Moreover, diverse classical problems deal just with operators of this type. However, one should not either lose sight of the limitations of these operators. For example, while they cover the Fejér means of the partial sums of the Fourier series of a periodic function, they do not encompass the partial sums themselves.

Before turning to these topics, which we shall do in the following chapter, we wish, by way of orientation toward the kind of results that can reasonably be expected, to see what light the general Banach space ("geometric") approach to approximation theory can shed upon degree-of-approximation questions. (The historical development was, of course, otherwise - first rather concrete theorems were found, and only later did more general theories attempt to absorb and "clarify" them.)

<u>8.2 Approximation of classes of functions by finite dimensional subspaces</u>. We shall here consider the case where X is a separable Banach space and the approximating classes are finite-dimensional subspaces X_n. The most important question is, for which closed subsets Y of X do we have

$$\lim_{n \to \infty} \operatorname{dev}(Y; X_n) = 0 \ ?$$

In preparation for the following theorem, let us recall from metric topology that <u>an</u> ϵ-<u>net</u> for a subset Y of a metric space (or, <u>a set which is</u> ϵ-<u>dense in</u> Y) is a set F such that every element of Y is at distance at most ϵ from F. By virtue of a basic theorem of Hausdorff, <u>a closed set</u> Y <u>is compact if and only if for every</u> $\epsilon > 0$ <u>there exists a finite</u> ϵ-<u>net for</u> Y. An immediate consequence is: if Y is a closed set which admits, for each $\epsilon > 0$, a <u>compact</u> ϵ-net, then Y is compact.

<u>8.2.1 Theorem</u>. <u>Let</u> X <u>be a Banach space and let</u> X_n, $n = 1, 2, \dots$, <u>be finite-dimen-</u>

sional subspaces satisfying $X_1 \subset X_2 \subset \ldots \subset X$, such that $\bigcup_{n=1}^{\infty} X_n$ is dense in X. Then, for a closed bounded subset Y of X

$$(1) \qquad \lim_{n \to \infty} \sup_{y \in Y} \operatorname{dist}(y, X_n) = 0$$

if and only if Y is compact.

Proof. (i) If Y is compact, there exists to each $\epsilon > 0$ a finite ϵ-net $y_1, \ldots, y_N$ for Y. By hypothesis there exist numbers n_i such that each y_i can be approximated within ϵ by some $x_i \in X_{n_i}$, so if $n_0 = \max n_i$, $i = 1, \ldots, N$, then X_{n_0} contains a point at distance at most ϵ from each of $y_1, \ldots, y_N$. Thus, for all $y \in Y$, $\operatorname{dist}(y, X_{n_0}) \leqslant 2\epsilon$; this implies (1).

(ii) Consider the sets $Y_i = \left\{x \in X_i : \|x\| \leqslant \rho + 1\right\}$, where $\rho = \sup_{y \in Y} \|y\|$. If (1) holds, then there exists, for each $\epsilon > 0$, $n = n(\epsilon)$ such that $\operatorname{dist}(y, Y_n) < \epsilon$, all $y \in Y$. But Y_n is compact, and is an ϵ-net for Y, which implies the compactness of Y. ◇

Thus, in the study of "degree of approximation" by finite-dimensional linear subspaces, it is natural to restrict attention to the approximation of compact sets of functions (this, by the way, is not the case for some typical non-linear approximating families, such as rational functions). Our next task will be to examine what the compact sets are in typical function spaces.

Exercise. Let Y denote the (non-linear!) subset of $C[0, 1]$ consisting of the functions $\left\{f_c\right\}$, $0 < c \leqslant 1$: $f_c(x) = c/(x + c)$. Show that Y contains an infinite subset of elements having mutual distances $\geqslant 1/2$. (The point is, even a (bounded) one-parameter family of rational functions may fail to admit a finite ϵ-net, so we certainly cannot expect a theorem like 8.2.1 for rational approximation.)

8.3 Compact sets in $C(Q)$. We take for granted the classical Arzelà-Ascoli theorem: if Q is a compact metric space, a closed bounded subset of $C(Q)$ is compact if and only if the functions comprising this subset are uniformly equicontinuous. This theorem can be reformulated, in a fashion more useful in approximation theory. Denoting by $d(\ ,\ )$ the distance function in Q, the modulus of continuity of $f \in C(Q)$

is defined as the function

$$\omega(a) = \omega(f;\, a) = \sup_{\substack{s,t \in Q \\ d(s,t) \leq a}} |f(s) - f(t)| \ .$$

Thus, ω is defined for $a \geq 0$. It is non-negative, increasing as a increases, and $\lim_{a \to 0} \omega(a) = 0$.

In terms of moduli of continuity, the Arzelà-Ascoli theorem can be stated: <u>the closed bounded set</u> $Y \subset C(Q)$ <u>is compact iff</u>

$$\lim_{a \to 0} \sup_{f \in Y} \omega(f;\, a) = 0 \ .$$

In other words, compact sets are characterized by the fact that the function

$$\omega(Y;\, a) \underset{\text{def.}}{=} \sup_{f \in Y} \omega(f;\, a)$$

tends to zero as $a \to 0$. It is clear that, in some sense, the rate of decrease to zero of $\omega(Y;\, a)$ is a measure of the "massivity" of Y, more rapid decrease indicating a less "massive" compact set Y, and one which ought to admit quantitatively better approximation theorems. In other words, one could reasonably expect that dev $(Y;\, X_n)$ should admit an estimate in terms of $\omega(Y;\, a_n)$ where a_n is some function of n (whose closer determination will depend on the X_n) satisfying $\lim_{n \to \infty} a_n = 0$; and, <u>a fortiori</u>, that for an individual $f \in C(Q)$, dist $(f,\, X_n)$ should admit an estimate in terms of $\omega(f;\, a_n)$. We shall see that these expectations are indeed borne out, for the special subspaces X_n which we study; the most fundamental <u>quantitative</u> versions of the Weierstrass approximation theorems, those of Jackson, fit the expected pattern exactly.

<u>8.3.1 A remark about the "entropy" concept</u>. Of course, we are here being wise after the fact. The object of the "abstract" theory of degree of approximation was not, however, to "anticipate" Jackson's theorems from the vantage point of hindsight, but to create a framework within which a large variety of problems should fall into place. The insight, due apparently to Kolmogorov, and pursued vigorously by him and his students, that <u>compactness</u> is the key concept in <u>linear</u> degree-of-approximation theory, has proven to be of great value. Moreover, it led to quantitative studies of

compactness in various concrete function spaces (e.g. the question of the minimum number of points in an ϵ-net) that have considerable intrinsic interest. The study of "entropy" in metric spaces (for the definition, cf. LORENTZ_1, TIMAN) originated here, and close ties between the theory of approximation and the theory of storage of information ("tables") were established. From the standpoint of classical approximation theory, the main use of these investigations is to provide lower bounds for the degree of approximation by function classes which may be of quite general character. The underlying principle is very simple: if we have the ("approximating") set S which is ϵ-dense in Y, and S is known to admit an ϵ-net consisting of N points, then clearly Y admits a 2ϵ-net consisting of N points. But suppose ϵ were so small that we could exhibit N + 1 points of Y which were at mutual distances $> 4\epsilon$ from one another - this would clearly contradict the preceding assertion, and in this way a lower bound is forced for ϵ, and consequently for dev (Y; S), in terms of geometric parameters belonging to Y and S. This argument can be refined in various ways.

To obtain upper bounds for the degree of approximation afforded by some particular class, e.g. polynomials of degree n, requires however some constructive moment geared to that particular approximating class; here the entropy concept is not likely to be helpful, and still less can it be of use where the approximation of individual functions is concerned. (For more about entropy, see KOLMOGOROV & TIHOMIROV, also survey articles of Lorentz referred to in LORENTZ_1.)

8.3.2 Remark. A slight reformulation of the Arzelà-Ascoli theorem is worth noting explicitly. If Ω denotes an increasing continuous function on $[0, \infty)$ satisfying $\Omega(0) = 0$, then the class

$$K^{\Omega} = \left\{f \in C(Q) : \omega(f; a) \leq \Omega(a)\right\}$$

is uniformly equicontinuous; hence the bounded closed subsets of K^{Ω} are compact; moreover, every compact subset of C(Q) is contained in some K^{Ω}. The classes K^{Ω}, which generalize the classical Lip α classes, play an important role in degree-of-approximation theorems. In the quantitative hierarchy of compact sets in C(Q), they are relatively "thick" or "massive" sets, the more so the more slowly $\Omega(a)$ tends to zero as $a \to 0$.

8.4 Compact sets in $L^p(T)$ and $L^p(R)$. In this paragraph, we assume $1 \leq p < \infty$, and denote as usual by T the unit circle. Let us recall our convention that we pass freely, whenever this is convenient, from one to the other interpretation of functions on T (i.e. either as functions on a circle, where integration is with respect to Haar measure of the circle, or as functions on R which have period 2π, integration being with respect to Haar measure of R).

8.4.1 Lemma (Approximate Identity). Let $k_n \in L^1(R)$, $n = 1, 2, \ldots$ satisfy

(i)
$$\int k_n(t)\, dt = 1, \qquad n = 1, 2, \ldots$$

(ii)
$$\sup_n \|k_n\|_1 < \infty$$

(iii)
$$\lim_{n \to \infty} \int_{|t| \geq \delta} |k_n(t)|\, dt = 0, \qquad \text{all } \delta > 0 .$$

Then, for each $f \in L^p(R)$, $1 \leq p < \infty$ the convolution

(1)
$$(f * k_n)(t) = \int_{-\infty}^{\infty} f(t - u)\, k_n(u)\, du$$

satisfies

(2)
$$\lim_{n \to \infty} \|f - (f * k_n)\|_p = 0$$

The same conclusion holds if $f \in L^p(T)$ (the convolution still being defined by (1)) provided that in (2) the norm $\| \;\|_p = \| \;\|_{L^p(R)}$ is replaced by $\| \;\|_{L^p(T)}$.

Proof. We leave this as an exercise. The lemma is true for a more general class of spaces (homogeneous Banach spaces on R^n with continuous translations, cf. next chapter). We also remark that the natural method of proof shows that (2) holds uniformly for f in compact subsets of $L^p(R)$ (and the analogous assertion for $L^p(T)$).

8.4.1.1 Exercises. a) Show that, if $h \in L^1(R)$ and $\int h\, dx = 1$, then $\{k_n\}$ where

$k_n(x) = nh(nx)$, is an approximate identity.

b) Prove Lemma 8.4.1, including the uniformity assertion.

c) Show that for f uniformly continuous and bounded on R, and $h \in L^1(R)$, $\int h\, dx = 1$ we have $\lim_{a \to 0} \int f(x - ay)\, h(y)\, dy = f(x)$, the convergence being uniform on all of R.

8.4.2 Theorem. Let Y be a closed bounded subset of $L^p(T)$, $1 \leq p < \infty$. Let T_u denote translation by u: $(T_u f)(t) = f(t - u)$. Then, Y is compact if and only if

$$(1) \qquad \lim_{\delta \to 0} \sup_{f \in Y} \|T_\delta f - f\|_p = 0 ,$$

i.e. the translations are uniformly equicontinuous in Y.

Proof. (i) If Y is compact there exists to given $\epsilon > 0$ a finite ϵ-net $f_1, \ldots, f_N$ in Y. Since translations are continuous in $L^p(T)$

$$\lim_{\delta \to 0} \|T_\delta f_i - f_i\|_p = 0 \qquad i = 1, \ldots, N ,$$

and (1) then follows easily from this with the aid of the triangle inequality.

(ii) Let $k_n(t) = nh(nt)$, $h(t) = \pi^{-1/2} e^{-t^2}$. Then k_n is an approximate identity. Consider the map $f \to f * k_n$ defined by 8.4.1(1). It is simple to verify, with the aid of (1), that the image Y_n of Y under this map is a closed, bounded, uniformly equicontinuous subset of C(T). Y_n is therefore compact as a subset of C(T), hence for each $\epsilon > 0$ admits a finite ϵ-net relative to the uniform norm, which is a fortiori an ϵ-net relative to the smaller $L^p(T)$ norm. For large n, by 8.4.1, and (1), Y_n is ϵ-dense in Y, and so Y admits a finite 2ϵ-net. ◇

In $L^p(R)$ we get substantially the same theorem, except that now also the "tails" of the functions in Y must be uniformly small. We state without proof

8.4.3 Theorem. A closed bounded subset Y of $L^p(R)$, where $1 \leq p < \infty$, is compact if and only if

$$(i) \qquad \lim_{\delta \to 0} \sup_{f \in Y} \|T_\delta f - f\|_p = 0 ,$$

(ii) $$\lim_{a \to \infty} \sup_{f \in Y} \int_{|t| \geq a} |f(t)|^p \, dt = 0 .$$

Analogously to the discussion in 8.3, we can reasonably expect the function

$$\omega_p(f; a) = \sup_{0 \leq \delta < a} \|T_\delta f - f\|_p$$

(the "L^p modulus of continuity of f") to play a role in degree-of-approximation theorems for $L^p(R)$ norm, and a similar remark applies to the corresponding $L^p(T)$ m.o.c..

<u>8.5 A negative theorem</u>. The "degree of approximation" problem, as we said earlier, involves estimating the numbers dev $(Y; X_n)$ as $n \to \infty$, where, typically, Y is a compact subset of the Banach space, and the X_n nested finite-dimensional subspaces whose union is dense. Under these conditions $\lim_{n \to \infty}$ dev $(Y; X_n) = 0$. Not surprisingly, we cannot assert more than this without further hypotheses about Y, not even if Y consists of a single element. This is the point of the "inertia theorem" 8.5.2 below - there are elements which are approximated arbitrarily badly. For its proof we require a lemma of F. Riesz on the existence of elements "almost orthogonal" to a subspace.

<u>8.5.1 Lemma (F. Riesz)</u>. <u>Let</u> X <u>be a normed linear space and</u> Y <u>a closed proper subspace</u>. <u>For given</u> $\varepsilon > 0$, <u>there exists an element</u> $x_\varepsilon \in X$ <u>with</u> $\|x_\varepsilon\| = 1$ <u>and</u> $\text{dist}(x_\varepsilon, Y) \geq 1 - \varepsilon$.

<u>Proof</u>. Let $x \in X \setminus Y$; since Y is closed, dist $(x, Y) = \delta > 0$. Given ε, $0 < \varepsilon < 1$, there exists $y_\varepsilon \in Y$ such that $\|x - y_\varepsilon\| \leq \delta/(1 - \varepsilon)$. We claim that $x_\varepsilon = (x-y_\varepsilon)/\|x-y_\varepsilon\|$ satisfies the requirements. Indeed, $\|x_\varepsilon\| = 1$ and for any $y \in Y$,

$$\|x_\varepsilon - y\| = \|(x - y_\varepsilon) - (\|x - y_\varepsilon\|)y\| \,/\, \|x - y_\varepsilon\| \geq \delta \,/\, \|x - y_\varepsilon\| \geq 1 - \varepsilon . \quad \diamond$$

<u>8.5.2 Theorem</u>. <u>Let</u> X <u>be any Banach space, and</u> $X_1, X_2, \ldots$ <u>any sequence of closed proper subspaces of</u> X. <u>Then, for any sequence of numbers</u> $\delta_n > 0$ <u>tending to zero</u>, <u>there exists an element</u> $x \in X$ <u>such that</u> dist $(x, X_n) \neq O(\delta_n)$.

Proof. Suppose, on the contrary, that dist $(x, X_n) = O(\delta_n)$ for every $x \in X$ and define

$$E_m = \left\{x \in X: \text{dist}\,(x, X_n) \leq m\,\delta_n,\ \ n = 1, 2 \ldots\right\}.$$

The sets E_m are closed and their union is X, so by Baire's theorem some E_{m_0} contains a ball B; that is, dist $(x, X_n) \leq m_0\,\delta_n$, $x \in B$, and all n. Since the ball $-B = \{-x: x \in B\}$ has the same property and E_{m_0} is convex, we can find a ball (which we shall continue to denote by B) with this same property, having its center at the origin. Let ρ denote the radius of B. If now n_0 is chosen so large that $m_0\,\delta_{n_0} \leq \rho/2$, every point of B is at distance $\leq \rho/2$ from X_{n_0}, which is a contradiction to Riesz' lemma. ◇

This theorem shows that there exist elements which are arbitrarily poorly approximable, e.g. there is a function in C(I) not approximable to within $O((\log \log n)^{-1})$ by polynomials of degree n, etc. In case the X_n are finite dimensional (which was not assumed in the above proof) a more precise theorem holds:

Let X be a Banach space, and $X_1, X_2, \ldots$ nested finite-dimensional subspaces whose union is dense in X. For any non-increasing sequence of positive numbers δ_n satisfying $\lim_{n \to \infty} \delta_n = 0$, there exists an element x of X satisfying dist $(x, X_n) = \delta_n$, $n = 1, 2, \ldots$.

This theorem was proved in a special setting by S. Bernstein, and adapted to Banach spaces by A.F. Timan, cf. TIMAN, p. 40.

8.6 Further "negative" results; the symmetrization technique and its applications

8.6.1 Generalities. Suppose we have a bounded linear operator A on the Banach space X. If we know in addition a compact group of operators acting on X, this fact can sometimes be exploited to get a lower bound for $\|A\|$ (the operator norm of A). For example, suppose G is a compact topological group with right Haar measure m, so normalized that m(G) = 1, and G is realizable as a group of isometries of X. That is, we have a group G′ of isometries of X, say $G' = \{I_g: g \in G\}$, which is topologically isomorphic to G in the natural way ($I_{g_1 g_2} = I_{g_1} I_{g_2}$, etc.). Then, we can define the symmetrization A_s of A (relative to G′) as the operator

$$A_s = \int (I_{g^{-1}} A I_g)\, dm(g) ,$$

the integration of the (operator valued, continuous) integrand causing no difficulties. Clearly $\|A_s\| \leq \|A\|$. Now, in concrete situations, A_s can often be computed explicitly, hence also $\|A_s\|$, which furnishes the desired lower bound on $\|A\|$. This technique seems first to have been employed by D.L. Berman to prove a minimal property of the Fourier projection; later Rudin applied it to obtain a new proof of a theorem of D.J. Newman on projections in $H^1(T)$. We shall now illustrate the technique in a concrete situation.

<u>8.6.2 Minimality of the Fourier projection, and applications</u>. Let F_n denote the <u>Fourier projection</u>, i.e. the operator on $L^1(T)$ which takes each f into the partial sum $s_n(t; f)$ of its Fourier series. Since $L^p(T) \subset L^1(T)$ for $p \geq 1$, we can also consider F_n as an (obviously, bounded) operator on $L^p(T)$. We already know that the $L^2 \to L^2$ norm of F_n is 1, since in fact F_n, as an operator in L^2, is simply orthogonal projection on $\mathcal{T}_n$. We now prove the striking theorem of BERMAN. Let us recall that a bounded linear operator A on the Banach space X is said to be a <u>projection on the subspace</u> X_o if its range lies in X_o and each element of X_o is left fixed by A.

<u>8.6.2.1 Theorem</u> (D.L. Berman). <u>Let</u> A <u>be any projection from</u> X <u>on</u> $\mathcal{T}_n$, <u>where</u> X <u>is one of the spaces</u> $L^p(T)$, $1 \leq p < \infty$ <u>or</u> C(T). <u>The operator norm of</u> A <u>is not less than that of the Fourier projection</u> F_n. <u>In particular, when</u> $X = C(T)$, $\|A\| \geq (4/\pi^2)\log n + O(1)$.

<u>Proof</u>. We consider the rotation group of the circle T as inducing isometries of X by the obvious change of variable: $(I_t f)(u) = f(u - t)$, $f \in X$, where t, u are elements of T (i.e. real numbers (mod 2π)). Let us compute the symmetrization A_s of A by computing its action on ψ_k, $\psi_k(t) = e^{ikt}$. Suppose first $|k| \leq n$. Then $(I_{-t}\, A\, I_t\, \psi_k)(u) = \psi_k(-t)(I_{-t}\, A\, \psi_k)(u) = \psi_k(-t)(I_{-t}\, \psi_k)(u) = \psi_k(u)$, hence, integrating with respect to t we see that $A_s\, \psi_k = \psi_k$. Consider next the case $|k| > n$. Again $(I_{-t}\, A\, I_t\, \psi_k)(u) = \psi_k(-t)(I_{-t}\, A\psi_k)(u)$ and, because A is a projection on $\mathcal{T}_n$, $A\psi_k = \varphi$ is a trigonometric polynomial of degree at most n; hence

$$(A_s \psi_k)(u) = (2\pi)^{-1} \int_0^{2\pi} (I_{-t} A I_t \psi_k)(u)\, dt = (2\pi)^{-1} \int_0^{2\pi} \psi_k(-t)\, \varphi(u+t)\, dt = 0$$

since ψ_k is L^2-orthogonal to the elements of $\mathcal{T}_n$. Thus $A_s \psi_k$ equals ψ_k or 0 according as $|k| \leq n$ or $|k| > n$. Therefore A_s, which is a bounded operator on X, agrees with the Fourier projection on the dense subset consisting of trigonometric polynomials, and so $A_s = F_n$.

Finally, it is a basic result of Fourier analysis that the $C(T) \to C(T)$ norm of F_n is the "Lebesgue constant"

$$\Lambda_n = (1/\pi) \int_0^{\pi} (|\sin (2n+1)t| / \sin t)\, dt = (4/\pi^2) \log n + O(1)$$

as $n \to \infty$, cf. ZYGMUND vol. 1. ◇

<u>Remarks</u>. a) The $L^1 \to L^1$ norm of F_n is also Λ_n (cf. Exercise 8.6.3 a) below). (It is known, and somewhat deeper, cf. ZYGMUND, that the $L^p \to L^p$ norm of F_n, for fixed $p > 1$, remains bounded as $n \to \infty$; in fact the Fourier series of $f \in L^p(T)$ converges in L^p norm (and even a.e.) to f.)

b) Recently CHENEY <u>et al</u> proved that the Fourier projection is the <u>unique</u> projection of minimal norm from C(T) to $\mathcal{T}_n$.

<u>8.6.2.2 Application to trigonometric interpolation</u>. The most striking application of the above theorem is to trigonometric interpolation. That is, let f be a given function in C(T) and suppose for each n we select a set E_n consisting of 2n + 1 points of the circle (distinct, or with multiplicities). Because $\mathcal{T}_n$ is a Haar (2n + 1)-space (cf. Chapter 2) there is then a unique trigonometric polynomial $t_n \in \mathcal{T}_n$ which interpolates f on the set E_n (at least, this is clear if the points of E_n are distinct; for a full discussion, including the modifications in the case of multiple points, cf. JACKSON, or ZYGMUND vol. 2). <u>Moreover, the map</u> $f \to t_n$ <u>is a projection from</u> C(T) <u>onto</u> $\mathcal{T}_n$, since when f belongs to $\mathcal{T}_n$ its interpolating polynomial agrees with it at 2n + 1 points and hence, by the Haar property, identically. Denoting by A_n the thus induced projection operator we have, by the preceding theorem,

$\lim_{n \to \infty} \|A_n\| = \infty$ and consequently: <u>there exists</u> $f \in C(T)$ <u>whose sequence of interpolating polynomials (based on the prescribed family of nodal sets</u> E_n) <u>fails to be uniformly bounded (and a fortiori fails to converge uniformly</u>).

Indeed, if the sequence $\{A_n f\}$ were bounded in $C(T)$, for every $f \in C(T)$, we should deduce the boundedness of $\|A_n\|$ by the Banach-Steinhaus theorem.

This negative result (and the analogous one discussed below concerning interpolation by algebraic polynomials) can be proved by more conventional methods of classical analysis, but the proofs are much more difficult. The method employed here is a striking example of the power of abstract analysis.

<u>8.6.2.3 Projections from</u> L^1 <u>onto</u> H^1. The same method of proof shows that <u>there is no bounded projectión from</u> $L^1(T)$ <u>to</u> $H^1(T)$ (Newman; cf. HOFFMAN, p. 154). Indeed, assuming A to be such a bounded projection, essentially the same computation as performed above shows that its symmetrization with respect to the rotation group of the circle is the "natural projection", i.e. the map N whose action on trigonometric polynomials is

$$N: \sum a_n e^{int} \to \sum_{n \geq 0} a_n e^{int} ,$$

but since N has no bounded extension to $L^1(T)$, the assumption that A was bounded leads to a contradiction. The same method of proof also establishes: <u>there is no bounded projection from</u> $C(T)$ <u>onto the disc algebra</u>.

<u>8.6.2.4 Generalizations</u>. The above argument obviously works in a more general setting: if G is any compact abelian group, and A any projection from $C(G)$ onto a closed translation-invariant subspace U of $C(G)$, the operator norm of A is not less than that of its symmetrization w.r.t. G, which is the "natural projection" N that leaves the characters in U fixed and maps the remaining characters into zero. (For related discussion see LAMBERT and references there.) This holds also in $L^p(G)$, and indeed, even in a wide class of Banach spaces of functions on G with translation-invariant norm (see the next chapter, where the case $G = R^n$ is discussed in detail).

In like manner, if $C(T)$ is replaced by $C(S_{n-1})$, where S_{n-1} is the unit sphere in R^n ($S_1 = T$), and H_m denotes the span of the spherical harmonics of degree not exceeding m (cf. MÜLLER), it is easily shown that the symmetrization w.r.t. the orthogonal group of any projection from $C(S_{n-1})$ onto H_m is the "natural projection", which coincides for "trigonometric polynomials" (finite linear combinations of spherical harmonics) with the orthogonal projection.

8.6.2.5 Application to algebraic polynomials. In order to apply the above technique to algebraic polynomials, it is convenient to consider a slightly modified situation: let $C^e(T)$ denote the even continuous functions of period 2π, and $\mathcal{T}_n{}^e$ its subspace of even ("cosine") trigonometric polynomials. Since $C^e(T)$ is not translation-invariant, we introduce the operators

$$S_t\colon (S_t f)(u) = (1/2)(f(u + t) + f(u - t))$$

which clearly map $C^e(T)$ into itself and have norm one. Suppose now A is a bounded projection from $C^e(T)$ to $\mathcal{T}_n{}^e$; then

$$A_\sigma = (1/2\pi) \int_0^{2\pi} (S_{-t}\, A\, S_t)\, dt$$

is a bounded linear operator on $C^e(T)$ and $\|A_\sigma\| = \|A\|$. To determine A_σ, we compute its action on $\psi_k(u) = \cos ku$. A computation like that above shows that

$$A_\sigma \psi_k = \begin{cases} \psi_0 , & k = 0 \\ \psi_k/2 , & 1 \leq k \leq n \\ 0 , & k > n \end{cases}$$

and from this it follows that $A_\sigma = (F_n + M)/2$ where M is the linear functional $Mf = (2\pi)^{-1} \int_0^{2\pi} f(t)\, dt$. Thus $\|A_\sigma\| \geq (\|F_n\| - 1)/2$ so that the $C(T) \to C(T)$ norm of A is not less than $(\Lambda_n - 1)/2 = (2/\pi^2) \log n + O(1)$.

An immediate application is the following result of Lozinski and Harshiladze (cf. NATANSON):

Let A denote any projection from $C(I)$, $I = [-1, 1]$ onto $\mathcal{P}_n$ (the "algebraic" polynomials of degree at most n). The $C(I) \to C(I)$ operator norm of A exceeds c log n, c = absolute constant.

For, the change of variables x = cos t induces an isometry of C(I) onto $C^e(T)$ which maps $\mathcal{P}_n$ onto $\mathcal{T}_n^{\,e}$, and hence induces a projection $\tilde{A}$ from $C^e(T)$ to $\mathcal{T}_n^{\,e}$, with the same norm as A, and to which the preceding can be applied.

Two important classes of projections are those induced by interpolation and orthogonal expansion. In the former case, a set $E_n \subset I$ consisting of n + 1 points is selected (n = 0, 1, ...) and the Lagrange interpolating polynomial corresponding to the given $f \in C(I)$ is formed. In the latter case, some bounded positive measure ρ on I is selected, and we expand f in its Fourier series with respect to the orthonormal polynomials belonging to ρ. In both cases, we have to do with projections on $\mathcal{P}_n$, and consequently there exists: (i) to each system $\{E_n\}$ of nodal points an $f \in C(I)$ whose associated Lagrange interpolating polynomials are not uniformly bounded, and (ii) to each ρ an $f \in C(I)$ the partial sums of whose Fourier expansion in ρ-orthogonal polynomials are not uniformly bounded.

Naturally, these results are far from exhausting the questions at hand, since they say nothing about pointwise behaviour, behaviour of subsequences, application of summability methods, behaviour in various weak topologies, etc., but they are none the less somewhat sobering, and show that projections have certain inherent weaknesses. In this connection, the following is of interest:

8.6.2.6 Theorem. Given $\epsilon > 0$, we can find a sequence of maps A_n from $X \to X$ (where X is $L^p(T)$, $1 \leq p < \infty$, or C(T)) such that

(i) The range of A_n lies in $\mathcal{T}_n$

(ii) A_n leaves each trigonometric polynomial of degree $\leq (1 - \epsilon)n$ fixed

(iii) The $X \to X$ operator norm of A_n satisfies $\|A_n\| \leq K \log (1/\epsilon)$ where K is an absolute constant

(iv) $\|A_n f - f\|_X \to 0$ as $n \to \infty$, all $f \in X$.

Proof. Let H_ϵ denote the "trapezoid function" equal to 1 on $[-1 + \epsilon, 1 - \epsilon]$, zero for $|x| \geq 1$ and piecewise linear, and $h_\epsilon \in L^1(R)$ the function whose Fourier transform is H_ϵ. Define, for $f \in X$

$$(A_n f)(t) = \int_{-\infty}^{\infty} f(t - (u/n))\, h_\varepsilon(u)\, du .$$

Then (recall 8.4.1.1 a)), $\lim_{n \to \infty} \|A_n f - f\| = 0$. Moreover, the Fourier coefficient of order k of $A_n f$ is $(A_n f)\hat{}(k) = f\hat{}(k)\, H_\varepsilon(k/n)$ which equals $f\hat{}(k)$ if $|k|/n \leqslant 1 - \epsilon$, and 0 if $|k| > n$, showing that (i) and (ii) hold. Finally, since A_n is the convolution of f with $nh_\varepsilon(nt)$ we have $\|A_n\| \leqslant \int_{-\infty}^{\infty} |h_\varepsilon(t)|\, dt < K \log (1/\epsilon)$ (the latter computation, we recall, was set as an exercise earlier, see exercise e), page VII-11).

◇

Thus, we can get "near-projections" which are uniformly bounded and have good convergence properties. We remark that here, as elsewhere in this chapter, we could take for X any "homogeneous Banach space", cf. next chapter.

In concluding this section, we may refer the reader to TIMAN and to the survey article $GOLOMB_2$ for supplementary details and references concerning the convergence properties of sequences of linear operators.

8.6.3 Exercises. a) Show that the $C(T) \to C(T)$ norm of the Fourier projection F_n equals Λ_n, and that $\Lambda_n >$ const. log n. Prove that the $L^1(T) \to L^1(T)$ norm of F_n is also Λ_n.

b) Prove that the "natural projection" (one-sided truncation) from $L^1(T)$ to $H^1(T)$ is unbounded, by producing trig. polynomials of norm one whose "natural projections" have large L^1 norms; do the same for the "natural projection" from C(T) to the disc algebra. Show that either of these results is deducible from the other.

c) Verify the assertions made in 8.6.2.4 that symmetrization of a projection is the "natural projection", both in the C(G) context and the $C(S_{n-1})$ context.

d) Estimate the $C(S_{n-1}) \to C(S_{n-1})$ norm (cf. 8.6.2.4) of the "natural projection" onto the span of the spherical harmonics of degree at most m.

e) Compute the interpolating trigonometric polynomials for $f \in C(T)$ when E_n consists of 2n + 1 equally spaced points on the circle, and show directly in this case that the norm of the corresponding projection is > const.log n.

f) Give a similar direct proof for the expansion of f into Chebyshev polynomials.

g) Prove that the map N in 8.6.2.3 has no bounded extension to $L^1(T)$.

8.7 Lower bounds for the degree of approximation via "widths"

8.7.1 Background and definitions. A very useful notion to "geometrize" the study of degree of approximation is that of the successive "widths" (or "diameters") of a compact set in a Banach space. Here we discuss only the barest details of the theory, referring the reader to TIHOMIROV$_1$, LORENTZ$_1$ or SINGER$_1$ for full details and further references. An n-plane in a Banach space X is defined to be any coset of an n-dimensional subspace (the latter considered as an additive subgroup of X); more concretely, an n-plane is the totality of elements $\{x_o + c_1x_1 + \dots c_nx_n\}$, where $x_i \in X$ are prescribed, $\{x_1, \dots, x_n\}$ being linearly independent, and $c_1, \dots c_n$ run through all scalars.

Let now K denote any subset of X. Then, the n-width of K (in X, we will say, if there is possible ambiguity as to the containing Banach space) is the (finite or infinite) quantity

$$(1) \qquad d_n(K) = \inf \operatorname{dev}(K; P_n)$$

as P_n ranges over all n-planes (we suppose here, of course, $n \leqslant \dim X$). Clearly we have

$$d_o(K) \geqslant d_1(K) \geqslant \dots \geqslant d_n(K) \geqslant \dots \quad .$$

In case an n-plane P_n^* exists such that $\operatorname{dev}(K; P_n^*) = d_n(K)$, P_n^* is said to be an optimal approximating n-plane to K.

For example, if K is an ellipsoid in ordinary Euclidean 3-space, with semi-axes $a_1 > a_2 > a_3$, its center is a (in fact, the unique) optimal approximating o-plane, and $d_o(K) = a_1$. The unique optimal approximating 1-plane is the line through the longest axis and $d_1(K) = a_2$. The unique optimal approximating 2-plane is the plane determined by the two longest axes, and $d_2(K) = a_3$.

In the general case, optimal approximating n-planes need not exist, i.e. the inf in (1) need not be attained. It is attained, however, for each bounded set K and each $n \geq 0$, *if* X *is a dual space*. This was proved by Garkavi; cf. SINGER$_1$, II, § 6.3 for this and further results. There may in general exist several optimal n-planes.

Let us assume henceforth, corresponding to the situations of greatest interest to us, that K *is symmetric about* 0 (i.e. $-x \in K$ whenever $x \in K$). This condition is readily seen to imply that in (1) we can restrict the P_n to be n-dimensional *subspaces* of X, and obtain the same infimum; in the case of an attained minimum, we speak of *an optimal* n-*dimensional approximating subspace* (for K). It is clear that, for K closed and bounded, $\lim_{n \to \infty} d_n(K) = 0$ if and only if K is compact.

8.7.2 Lower bounds for widths

8.7.2.1 The main utility of the width concept is that for many sets K of practical interest one can obtain *lower bounds* for $d_n(K)$ on the basis of fairly simple and general arguments. One has then a target to aim at, a lower threshold, when seeking to obtain direct (or "positive") theorems for the class K using as the approximating class one or another concrete linear family, e.g. ordinary polynomials. Let us illustrate this for subsets K of C(Q) with the following simple proposition (LORENTZ$_1$, p. 133), which we state without proof:

(*) *Let* Q *be a compact Hausdorff space, and* $K \subset C(Q)$. *Assume there exist* $n + 1$ *points* t_i $(i = 0, 1, \ldots, n)$ *of* Q *and a number* $\epsilon > 0$ *with the following property: for each choice of signs* $\sigma_i = \pm 1$, $i = 0, 1, \ldots, n$, *there is a function* $f \in K$ *such that*

$$\operatorname{sgn} f(t_i) = \sigma_i \ , \quad |f(t_i)| \geq \epsilon \ , \quad i = 0, 1, \ldots, n \ .$$

Then, the n-*width of* K *in* C(Q) *is not less than* ϵ.

8.7.2.2 For example, choosing $Q = [0, 1]$, and for K the normalized Lip 1 class:

$$K = \left\{ f \in C(Q) : |f(t_2) - f(t_1)| \leq |t_2 - t_1| \ , \quad \text{all} \ \ t_1, t_2 \in Q \right\} \ ;$$

we can choose $t_i = i/n$, $i = 0, 1, \ldots n$, and $\epsilon = 1/2n$. The hypotheses of the above proposition are satisfied, since the "broken line" function taking at t_i the value $\sigma_i/2n$ belongs to K. *Therefore*, dev $(K; P_n) \geq 1/2n$ *for any* n-*dimensional subspace* P_n

of $C[0, 1]$. Since, as we shall see, dev $(K, \mathcal{P}_{n-1}) \leq c/n$ for some absolute constant c, we can then conclude (i) that $d_n(K)$ has the order of magnitude $1/n$, and (ii) insofar as approximation of the class K is concerned, ordinary (algebraic) polynomials are substatially as effective as any other linear approximating family containing the same number of free parameters.

8.7.2.3 The latter conclusion, of obvious practical importance, is very typical for the results achieved via width estimates. When X is $C(Q)$ or $L^p(Q)$, Q lying in a Euclidean space, and K is a set defined e.g. by a modulus of continuity condition or generalization thereof, the lower bounds for $d_n(K)$ arrived at by quite general considerations have turned out to be fairly well matched by the upper bounds obtained from suitably constructed polynomial approximations. (For further elaboration of these remarks, see NEWMAN & SHAPIRO$_2$.)

8.7.2.4 Here, several qualifying remarks should be added. First of all, there is a certain intrinsic interest in the exact value of $d_n(K)$ (or at least, an asymptotic expression for it) and the (eventual) optimal n-dimensional approximating subspaces. This of course requires a recondite analysis in each particular case, and only quite limited results have been obtained (cf. TIHOMIROV$_1$). The famous success story here is that of the sets $K_r = \{f \in C(T): |f^{(r)}(t)| \leq 1\}$, where the exact values of $d_n(K_r)$ are known, and for example, $\mathcal{T}_m$ furnishes an optimal approximating subspace when $n = 2m + 1$ (precise formulation in TIHOMIROV$_{1,2}$; the demonstration of these results depends heavily on the theorem of Favard-Ahiezer-Krein which we proved in the preceding chapter (Theorem 7.5.2)).

8.7.2.5 Secondly, the "width" technique is restricted to linear approximating families. (The Soviet school has relied on entropy estimates to get lower bounds relative to non-linear approximating families.) However, for a large class of non-linear approximating families, one can adapt the techniques used by Kolmogorov, Lorentz, et al, in obtaining lower bounds for the widths. Indeed, these techniques are applicable whenever the approximating family S is what we might call oscillation-limited. Let us here, very sketchily, try to explain this idea. Consider first the case where S is an n-dimensional subspace of (real) $C(Q)$, write $N = n + 1$, and let $t_1, \dots, t_N$ be points of Q; then the following proposition is true:

(**) The set of sign patterns

$$\left\{\operatorname{sgn} f(t_1), \operatorname{sgn} f(t_2), \dots, \operatorname{sgn} f(t_N)\right\} \tag{1}$$

cannot exhaust all of the 2^N possible N-tuples of ± 1, as f ranges over S.

Indeed, this is an immediate consequence of the existence of real numbers c_i, not all zero, such that

$$\sum_{i=1}^{N} c_i f(t_i) = 0 , \qquad \text{all } f \in S . \tag{2}$$

Proposition (**) is the key to the proof of 8.7.2.1 (*).

Now, there are many non-linear families S depending on n parameters, for which (**) remains valid, with the modification, however, that N is not n + 1 but of the same order of magnitude. For example, if S is the class

$$(PP) \underset{\text{def.}}{=} \left\{fg\colon f \in P , \ g \in P\right\}$$

where P is an m-dimensional subspace (this class depends essentially on $n = 2m - 1$ parameters), (**) is valid with $N = 16\,m = 8(n + 1)$. (This is considerably more difficult to prove than the linear analog (since nothing like (2) is available in the non-linear case), and is based upon combinatory results in geometry, concerning the partitioning of Euclidean m-space by a collection of hyperplanes (see SHAPIRO$_8$)). Using this, one can establish 8.7.2.1 (*) with $d_n(K)$ replaced by dev (K, PP) and the n + 1 points t_i replaced by a set of 8(n + 1) points, and this enables one to show that the (non-linear) family (PP) is essentially no better for purposes of approximating K^Ω (cf. 8.3.2) than a linear family based on the same number of parameters. Hugh Warren, in his University of Michigan doctorate dissertation, further developed this method so as to handle non-linear families which depend polynomially or rationally on a finite number of parameters (WARREN). He thus derived afresh, and with sharper estimates, very difficult results which VITUSHKIN had obtained using "entropy" and "variations of sets" techniques. Warren required neither of these tools, but the price that had to be paid was that the proof of (**) with suitably small N now required difficult new estimates concerning the partitioning of Euclidean spaces by algebraic surfaces. These latter estimates have an independent interest, however, and represent an important contribution to real algebraic geometry not less than to approximation theory.

The idea embodied in (**) is by no means confined to $C(Q)$; in other Banach spaces one must, however, replace the point evaluations $f(t_i)$ appearing in (1) by appropriate linear functionals. In this way, Warren obtained also L^p versions of Vitushkin's results.

To summarize: non-linearity does not seem to be too serious an obstacle to obtaining lower bounds for the degree of approximation to "massive" sets, of the type of the sets K^Ω. The general trend of what has been learned so far is that these sets are not essentially better approximable by typical non-linear families than by linear ones, or even (in the classical context of the unit interval) by ordinary polynomials, depending on the same number of parameters. And the reason is: the non-linear classes S in question are "oscillation-limited" in the sense that (**) holds, with $N = N(n) = O(n)$. This limits the capacity of S to approximate any class K such that, for some $\varepsilon > 0$, the set of restrictions of its elements to some set $\{t_1,\dots,t_N\}$ contains all of the 2^N sequences $\{\pm\varepsilon, \pm\varepsilon, \dots, \pm\varepsilon\}$; for obviously then dev $(K;S) \geq \varepsilon$. (We are being very concise here, but must refer to the cited literature for further details).

8.7.2.6 However, for several classes not of the type K^Ω (e.g. classes of functions on I defined by bounds on higher-order derivatives, or subjected to restrictions of monotonicity or convexity) the method of 8.7.2.1 for estimating widths from below fails. Thus, faced with such a function class, we are in general very hard put to establish lower bounds for the degree of approximation by means of polynomials, rational functions, or anything else, and *ad hoc* devices must be created for each particular case. Thus, D.J. Newman not long ago made the startling discovery that each broken-line function is uniformly approximable on $[-1,1]$, by rational functions of degree n, with an error $O(e^{-c\sqrt{n}})$, and Szüsz and Turán extended this to piecewise holomorphic functions.

This is in contrast to the situation vis à vis (ordinary) polynomials: a broken-line function on $[-1, 1]$ can be uniformly approximated by an element of $\mathcal{P}_n$ with an error $O(1/n)$, but no better. The lower bound, i.e. the fact that $O(1/n)$ is the best possible, is easy to establish on the basis of the principles set forth in Chapter 2, since one can explicitly construct enough measures to annihilate $\mathcal{P}_n$. (For this reason, one can generally speaking establish a sharp lower bound for the degree of approximation by polynomials of a concretely given function. With rational approximation, however, there is nothing corresponding to duality theory, annihilating measures, etc., and the task is much more difficult.)

Following the work of Newman and Szüsz-Turán a series of results have been obtained, by Freud, Gonchar, Szabados and others showing that various natural function-classes are essentially better approximable by rational functions than by polynomials of the same degree. Szüsz and Turán had established such a result for convex functions. Some of the relevant papers are FREUD_2 (which contains references to the papers of Newman, Turán and Szüsz), $\text{GONCHAR}_{1,2}$, FREUD & SZABADOS, SZABADOS. The field is still developing intensively at the present time (WALSH_2).

Exercise. Show that there exists a constant $c > 0$, such that the function $|x|$ on I has distance $\geqslant c/(n + 1)$ from $\mathcal{P}_n$.

8.7.2.7 We should also mention that in $C(Q)$ and $L^p(Q)$, where Q is a ball or cube in a Euclidean space of many dimensions, although good lower estimates for $d_n(K)$ are known when K is e.g. a Lipschitz class, the analogous estimates present difficulties for various important "higher smoothness" classes e.g. Sobolev spaces and similar classes determined by bounds on the partial derivatives. We refer the reader to MITYAGIN, $\text{JEROME}_{1,2}$, and JEROME & SCHUMAKER for some results along these lines. In our opinion, the construction of nearly optimal polynomial and trigonometric polynomial approximations for some of these "higher smoothness" classes, along lines we have indicated in SHAPIRO_3, is beginning to look feasible, and further width estimates from below will be needed to go hand in hand with this development, as a check that one is really getting sharp results. (For polynomial approximation in several variables, not dealing however with the "width" estimates, see NIKOLSKI.)

8.7.2.8 In reviewing recent work related to widths, we should mention here also the pioneering research by D.J. Newman and his students on approximating the normalized Lipschitz 1 class K by (ordinary) polynomials in the C(Q) norm, where Q is now a "thin" subset of a Euclidean space. The results are still fragmentary, the general trend being that (in style with "thin set" problems elsewhere, e.g. in Fourier analysis) subtle metric properties of Q determine whether the polynomials of fixed degree are relatively near or relatively far from being optimal approximating subspaces for K (see NEWMAN, NEWMAN & $\text{RAYMON}_{1,2}$).

8.7.3 Widths in Hilbert spaces. In Kolmogorov's original investigation, X was a Hilbert space, and in this case the problem of widths (at least, for a special kind of compact set) is intimately related to an eigenvalue problem. Suppose, namely, H is a Hilbert space and A a compact (= completely continuous) linear operator on H, and denote by K the image under A of the unit ball. Thus the closure K^- of K is compact. Clearly K is symmetric about 0, so we may use subspaces (rather than planes) in computing the widths. We have

$$d_n(K) = \inf_{H_n} \operatorname{dev}(K, H_n)$$

where the inf is over all n-dimensional subspaces H_n of H. Let $\lambda_1 \geqslant \lambda_2 \geqslant \ldots \geqslant \lambda_n \geqslant \ldots \to 0$ denote the (not necessarily distinct) eigenvalues of the (Hermitian, compact, positive) operator $B = A^*A$. We have then

8.7.3.1 Theorem. $d_n(K) \geqslant \lambda_{n+1}^{1/2}$, and if A is Hermitian equality is attained, n = 0, 1, 2, ...

Proof. We have, denoting by S^o the orthocomplement of a subspace S of H, and by H_n a variable subspace of dimension n,

$$d_n(K)^2 = \inf_{H_n} \sup_{x \in K} \operatorname{dist}(x, H_n)^2 = \inf_{H_n} \sup_{\|y\| \leqslant 1} \inf_{z \in H_n} \|Ay - z\|^2$$

$$\geq \inf_{H_n} \sup_{\substack{\|y\| \leq 1 \\ Ay \in H_n^o}} \inf_{z \in H_n} \|Ay - z\|^2 = \inf_{H_n} \sup_{\substack{\|y\| \leq 1 \\ Ay \in H_n^o}} \inf_{z \in H_n} (\|Ay\|^2 + \|z\|^2)$$

$$= \inf_{H_n} \sup_{\substack{\|y\| \leq 1 \\ Ay \in H_n^o}} \|Ay\|^2 = \inf_{H_n} \sup_{\substack{\|y\| = 1 \\ y \in (A^* H_n)^o}} (By, y) \quad .$$

Since $A^* H_n$ has dimension at most n, we have therefore,

$$d_n(K)^2 \geq \inf_{0 \leq r \leq n} \inf_{H_r} \sup_{\substack{\|y\| = 1 \\ y \in H_r^o}} (By, y)$$

and by the classical Fischer minimax property of the eigenvalues,

$$d_n(K)^2 \geq \inf_{0 \leq r \leq n} \lambda_{r+1} = \lambda_{n+1} \quad .$$

Suppose now that A is Hermitian. Let $u_1, u_2, \ldots$ denote the (orthonormal) system of eigenvectors of A corresponding to the eigenvalues $\alpha_1, \alpha_2, \ldots$, the latter so ordered that $\alpha_n^2 = \lambda_n$, and denote by S_n the span of $\{u_1, \ldots, u_n\}$. Then,

$$d_n(K)^2 = \inf_{H_n} \sup_{\|y\| \leq 1} \inf_{z \in H_n} \|Ay - z\|^2 \leq \sup_{\|y\| \leq 1} \inf_{z \in S_n} \|Ay - z\|^2 \quad .$$

Now, writing $y = \sum_{m=1}^{\infty} c_m u_m$, where $\sum |c_m|^2 \leq 1$, we have

$$\inf_{z \in S_n} \|Ay - z\|^2 = \inf_{z \in S_n} \|z - \sum_{m=1}^{\infty} c_m \alpha_m u_m\|^2 = \sum_{m=n+1}^{\infty} |c_m|^2 \alpha_m^2 \quad ,$$

and so

$$d_n(K)^2 \leq \sup_{(c) : \sum |c_i|^2 = 1} \sum_{m=n+1}^{\infty} \alpha_m^2 |c_m|^2 = \alpha_{n+1}^2 = \lambda_{n+1}$$

completing the proof. ◇

Remarks. a) In the case where A is Hermitian, we see from the proof that the span of the eigenvectors $u_1, u_2, \ldots, u_n$ of A is an optimal approximating subspace of dimension n to the set K. For the unicity of this optimal subspace, cf. STENGER.

b) For related results see GOLOMB$_3$, JEROME$_1$.

One can use this result to compute the widths, in the Hilbert space $L^2(T)$, of various compact function classes characterized by bounds on quadratic functionals involving the functions and their derivatives (cf. LORENTZ$_1$, p. 143 ff, JEROME$_{1,2}$), or at least to reduce such problems to eigenvalue problems concerning which quite detailed results are available in the literature. Here we will only give a trivial illustration. Let H denote that subclass of $L^2(T)$ consisting of functions with mean value zero, and $A = A_r$ (r = positive integer) the (Hermitian) operator on H whose action on a function f with Fourier coefficients $\{f^\wedge(j)\}_{j=-\infty}^{\infty}$ is to transform it into the function with Fourier coefficients $\{j^{-r} f^\wedge(j)\}_{j=-\infty}^{\infty}$. Thus, A_r is (apart from a constant factor of modulus one) an "integrating operator", which maps f onto an r-fold primitive. In this case

$$K = \left\{f : (1/2\pi) \int_0^{2\pi} |f^{(r)} (e^{it})|^2 \, dt \leqslant 1\right\} ,$$

whereby f is supposed to have a continuous derivative of order r - 1, which is in turn the integral of an element (called $f^{(r)}$) of $L^2(T)$. The eigenvalues of A are $\pm (1^{-r}), \pm (2^{-r}), \ldots$; hence $d_{2n}(K) = (n + 1)^{-r}$. (This result can of course be proved out of hand without recourse to the preceding theorem). This is the L^2 analog of the problem solved for C(T) by the Favard-Ahiezer-Krein-Tihomirov theorem that we spoke of in 8.7.2.4.

Exercises. a) What are the conditions on the positive numbers λ_n in order that the "ellipsoid" $\sum_{n=1}^{\infty} x_n^2 / \lambda_n^2 = 1$ be compact in sequential Hilbert space? What are its widths?

b) What are the conditions on the positive numbers ρ_i in order that $\{x : |x_i| \leqslant \rho_i,\ i = 1, 2, \ldots\}$ be compact in sequential Hilbert space? What are the widths of this set?

c) Similar problem for $\left\{x: \sum \rho_i |x_i|^\alpha \leq 1\right\}$, where $\alpha > 0$.

d) Try the analogous problems in the Banach space of real sequences summable to the power p (real ℓ^p, $p \geq 1$).

<u>8.8 Generating good approximations by positive linear operators</u>. Let us now turn to some results of a positive character, showing that good approximations exist and how to construct them. A variety of constructive procedures are available, for various kinds of approximating classes such as polynomials, piecewise polynomials ("splines"), trigonometric polynomials, orthogonal systems, etc. Most of these procedures are <u>linear</u>, i.e. construct the "good" approximation to a given function by applying a suitable linear operator to it. (Since, except in Hilbert spaces, the mapping from an element to its best approximation from a subspace is in general <u>non-linear</u>, it is clear that linear methods can only give "good", and not "best" approximations.) Concerning the most classical situation, approximation by ordinary ("algebraic") polynomials on an interval, the main linear methods employed have been interpolatory schemes of various sorts, expansions in terms of orthogonal polynomials, and what may loosely be termed "smoothing" operations, either based upon a partition of unity, as with the Bernstein polynomials, or upon convolution with a suitable kernel. In these lectures we shall concentrate mainly on convolution operators, which are very effective in generating good approximations and offer advantages of both conceptual and computational simplicity, as well as flexibility in the choice of kernels. A possible drawback for numerical work might be that the approximating polynomial is given only implicitly, as a convolution integral, so that its coefficients are explicitly obtainable only after further computations. In 8.8.2.5 we shall show how convolution with a polynomial kernel can be used to generate good polynomial approximations to a continuous function on a compact set in Euclidean space. The search for a good kernel leads to an extremal "moment" problem for non-negative polynomials. First, however, we shall discuss a more instructive abstract approach to the same problem, the Bohman-Korovkin theory of positive operators.

<u>8.8.1</u> To fix ideas, let us look at the question of <u>uniform</u> approximation to a

continuous function f on I = [- 1, 1] by polynomials of degree n. We are now taking the "degree of approximation" standpoint, so we want a procedure which will give good results for large n, at the very least yield the Weierstrass theorem that dist $(f, \mathcal{P}_n) \to 0$ as $n \to \infty$. We have already seen that the Chebyshev line of attack _via_ "best approximation" is difficult to exploit here. Likewise, we know from material earlier in this chapter (and what Runge showed directly by a famous counterexample) that a procedure based on Lagrange interpolation _in its most naive form_ cannot work; nor can expansion in orthogonal polynomials. Concerning the approximation procedures which _have_ been found to be universally effective, many of these can be subsumed in a single general class ("positive operators"), leading to a unified theory and the elimination of widespread duplications in the traditional literature. We give here a brief account of this approach, the main ideas of which are due to Bohman and Korovkin, following NEWMAN & SHAPIRO$_2$.

8.8.2 Positive operators. If Q is a compact metric space, a linear operator A from (real-valued) C(Q) to itself is _positive_ if

$$f(x) \geqslant 0, \text{ all } x \in Q \Rightarrow (Af)(x) \geqslant 0, \text{ all } x \in Q .$$

An immediate consequence of the definition is: if A is a positive linear operator and $|f(x)| \leqslant g(x)$, then $|(Af)(x)| \leqslant (Ag)(x)$, so A is bounded.

Let d(,) denote the distance function in Q. We are interested in spaces Q which have the property

(*) _Whenever_ $d(x, y) = a + b$, _where_ $a > 0$, $b > 0$ _there exists a point_ $z \in Q$ _such that_ $d(x, z) = a$ _and_ $d(z, y) = b$.

Finally, let us denote by δ_y the function $x \to d(x, y)$, and by $\underline{1}$ the function identically equal to 1. The basis of the Bohman-Korovkin theory is

8.8.2.1 Lemma. If Q _satisfies_ (*), A _is a positive linear operator on_ C(Q) _satisfying_ $A\,\underline{1} = \underline{1}$, _and_ $f \in C(Q)$ _has modulus of continuity_ ω, _then_

$$(1) \qquad |(Af)(y) - f(y)| \leqslant \omega(t)[1 + t^{-1}(A\delta_y)(y)] \leqslant \omega(t)[1 + t^{-1}((A\delta_y^2)(y))^{1/2}]$$

for all $y \in Q$ _and_ $t > 0$.

Proof. Because of (*), for every $x, y \in Q$ there exist points $z_1, \ldots, z_m$, where $m \leqslant d(x, y)/t$, such that the distance between any two points in the chain $x, z_1, \ldots, z_m, y$ is at most t. Now, $|f(x) - f(y)| \leqslant (1 + t^{-1} d(x, y))\, \omega(t)$. Holding y fixed, and applying A to this inequality gives

$$|(Af)(x) - f(y)| \leqslant [1 + t^{-1}(A\delta_y)(x)]\, \omega(t) .$$

Setting $x = y$ gives the first inequality in (1). Now, the map $g \to (Ag)(z)$ is, for each fixed $z \in Q$, a continuous positive linear functional, (call it Λ_z) on $C(Q)$, and $\Lambda_z \underline{1} = 1$. Hence, as the usual proof of the Schwartz inequality shows, $(\Lambda_z g)^2 \leqslant \Lambda_z(g^2)$. Choosing here $z = y$ and $g = \delta_y$ gives $((A\delta_y)(y))^2 \leqslant (A\delta_y^2)(y)$ which yields the second inequality in (1). ◇

As an application, we have the celebrated theorem of Bohman-Korovkin.

8.8.2.2 Theorem. Let $\{A_n\}_{n=1}^{\infty}$ denote a sequence of positive linear operators from $C[a, b]$ to itself satisfying

$$\lim_{n \to \infty} A_n f = f \tag{1}$$

for each of the following three functions $f \in C[a, b]$: $f(x) = \underline{1}$, $f(x) = x$, $f(x) = x^2$. Then (1) holds for every $f \in C[a, b]$.

This theorem is very striking (and very useful) because (1) need only be checked for three "test functions" (which is usually trivial to do in practice) in order to conclude that it is universally valid.

Proof of theorem. In order to be able to quote the lemma, let us prove the theorem under the slightly stronger hypothesis that $A_n \underline{1} = \underline{1}$ for all n (with the weaker hypothesis $A_n \underline{1} \to \underline{1}$ only a trivial modification is needed). Clearly (*) is satisfied, so we can apply the lemma. Here δ_y^2 is the function $x \to (x - y)^2$ so writing $\varphi_i(x)$ for x^i $(i = 0, 1, 2)$ we have

$$(A_n\, \delta_y^2)(y) = (A_n\, \varphi_2)(y) - 2y(A_n\, \varphi_1)(y) + y^2(A_n\, \varphi_o)(y)$$

which, by hypothesis, converges (uniformly on $[a, b]$) to

$$\varphi_2(y) - 2y\,\varphi_1(y) + y^2\,\varphi_o(y) = 0 .$$

Therefore, an application of the lemma gives, for each $f \in C[a, b]$ and uniformly for $y \in [a, b]$,

$$\overline{\lim_{n\to\infty}}\ |(A_n f)(y) - f(y)| \leq \omega(t)[1 + t^{-1}\ \overline{\lim_{n\to\infty}}\ ((A_n\,\delta_y^2)(y))^{1/2}] = \omega(t) .$$

Here $t > 0$ is at our disposal, so that $\omega(t)$ can be made as small as we like. Therefore, the $\overline{\lim}$ is zero. ◇

Similar theorems can be established, by the same technique, for trigonometric approximation, polynomial approximation in several variables, etc. Cf. the elegant book of KOROVKIN (unfortunately not available in an accurate English translation). The theory has been further developed and refined in many recent papers, e.g. by Baskakov and by Shashkin (references in LORENTZ_2), by FREUD & KNAPOWSKI, et al.

8.8.2.3 Let us apply the theorem to the Bernstein polynomials. Here $[a, b]$ is $[0,1]$ and

$$(A_n f)(x) = B_n(f; x) \underset{\text{def.}}{=} \sum_{k=0}^{n} f(k/n)\,\beta_{n,k}(x); \qquad \beta_{n,k}(x) = \binom{n}{k} x^k (1 - x)^{n-k} .$$

By simple computations (which can be interpreted probabilistically as computing moments of a binomial distribution) (recall that $\varphi_i(x)$ denotes x^i):

$$B_n(\varphi_i; x) = \begin{cases} 1 & , \ i = 0 \\ x & , \ i = 1 \\ x^2 + n^{-1}\,x(1 - x) & , \ i = 2 . \end{cases}$$

Thus, $\lim_{n\to\infty} \|A_n\,\varphi_i - \varphi_i\| = 0$, $i = 0, 1, 2$ and by 8.8.2.2 we conclude that $A_n f \to f$ for every $f \in C[0, 1]$, i.e. the Bernstein polynomials $B_n(f; x)$ of a continuous function f on $[0, 1]$ converge uniformly to f. Since $B_n \in \mathcal{P}_n$, the Weierstrass polynomial approximation theorem is a corollary.

8.8.2.4 Degree of approximation. Let us define (in the context of 8.8.2.1)

$$(1) \qquad \rho(A) = \max_{y \in Q}\ (A\delta_y)(y) .$$

In view of the Schwartz inequality, we have also

(2) $$\rho(A) \leqslant \max_{y \in Q} ((A\,\delta_y^2)(y))^{1/2} .$$

Lemma 8.8.2.1 has as a consequence the estimate

(3) $$\|Af - f\| \leqslant \inf_{t > 0} \omega(t)[1 + (\rho(A)/t)] \leqslant 2\,\omega(\rho(A)) .$$

These relations enable us to estimate the error of approximation in terms of the modulus of continuity ω. For example, when A_n is the above defined "Bernstein operator", $(A_n\,\delta_y^2)(y) = n^{-1}\,y(1-y) \leqslant 1/4n$. Hence, from (3),

$$|B_n(f;\,x) - f(x)| \leqslant \|A_n f - f\| \leqslant 2\,\omega(\rho(A_n)) \leqslant 2\,\omega((4n)^{-1/2}) .$$

Thus, the Bernstein polynomials approximate with an error of order not exceeding $\omega(n^{-1/2})$. It may be shown by examples that Bernstein polynomials in general need not approximate a function with an error of smaller order than this (and moreover, they cannot approximate to any function whatsoever, that is not a first degree polynomial, with an error that is $o(1/n)$ - "saturation phenomenon"). This behaviour is far from optimal, since we can achieve the order $\omega(n^{-1})$ by a suitably constructed polynomial, as we shall see. Thus, from a pure "degree of approximation" standpoint, Bernstein polynomials are not especially interesting. They have, however, many other remarkable, compensating qualities, e.g. they oscillate no more than the function which gives rise to them (as measured by total variation), display interesting properties when considered in the complex plane, etc. For a detailed study see LORENTZ_3.

It is not hard to achieve the $\omega(n^{-1})$ degree of approximation by means of a suitable sequence of positive operators $A_n : C(I) \to \mathcal{P}_n$, cf. NEWMAN & SHAPIRO_2, BOJANIC & DeVORE; the "trigonometric" version is in KOROVKIN. At this point, however, we wish to pass to a brief discussion of the use of convolution integrals as such. Once again we emphasize that we have in no way aimed at completeness in the preceding sections of this chapter. The reader interested in deeper study should consult, in addition to the sources we have mentioned, the Soviet literature especially, where most of the topics of this chapter have been studied intensively. Useful guides to this literature are the review articles $BUCK_3$, $LORENTZ_2$ as well as TIMAN, the (already cited) survey articles $TIHOMIROV_1$, KOLMOGOROV & TIHOMIROV, and KUPTSOV, and several proceedings (not available in English) of conferences devoted to "constructive theory of functions".

<u>8.8.2.5 Convolution with a polynomial kernel - Jackson's theorem</u>. Let K denote a compact set in R^m which contains the origin O. Let f denote a continuous function on K. <u>We suppose that</u> f <u>vanishes on the boundary of</u> K. If $\tilde{f}$ denotes the function equal to f on K and zero outside, then it is trivial to verify that $\tilde{f}$ has the same modulus of continuity $\omega(a) = \omega(\tilde{f}; a)$ as f. Let now $\mathcal{P}_n = \mathcal{P}_n^m$ denote the class of polynomials of degree at most n in the m variables $t_1, \dots t_m$, i.e. the span of those monomials $t_1^{n_1} \dots t_1^{n_m}$ for which $n_1 + \dots + n_m \leqslant n$. Suppose now $p \in \mathcal{P}_n$, then

$$(1) \qquad q(t) = \int_{R^m} \tilde{f}(t-u)\, p(u)\, du = \int_{R^m} \tilde{f}(u)\, p(t-u)\, du$$

is evidently an element of $\mathcal{P}_n$ (there is no convergence problem because $\tilde{f}$ has compact support). Now, <u>suppose</u> t <u>is restricted to lie in</u> K. Then clearly $\tilde{f}(t-u) = 0$ unless $t - u$ also lies in K, i.e. $u \in t - K = \{t - t' : t' \in K\}$. Let

$$K - K \underset{\text{def.}}{=} \{t' - t'' : t' \in K,\ t'' \in K\} .$$

Let us suppose now

$$(2) \qquad \int_{K-K} p(t)\, dt = 1 .$$

Then, for $t \in K$

$$|f(t) - q(t)| = \left| \int_{K-K} (\tilde{f}(t) - \tilde{f}(t-u))\, p(u)\, du \right| \leqslant \int_{K-K} \omega(|u|)\, |p(u)|\, du$$

This estimate is often useful in constructing polynomial approximations to f. For example, if λ is any positive number, we get the estimate

$$|f(t) - q(t)| \leqslant \omega(\lambda) \int_{K-K} (1 + |u|/\lambda)\, |p(u)|\, du = \omega(\lambda)(A + B\lambda^{-1}) ,$$

where

$$A = \int_{K-K} |p(u)|\, du \quad , \quad B = \int_{K-K} |u| \cdot |p(u)|\, du \quad .$$

Moreover, $A = 1$ (by (2)) if $p(t) \geqslant 0$ on $K - K$. We get therefore (choosing $\lambda = B$ with B minimal): *If* $f \in C(K)$, *there exists* $q \in \mathcal{P}_n$ *such that*

$$|f(t) - q(t)| \leqslant 2\, \omega(f;\, \lambda_n) \ , \quad t \in K$$

where λ_n *denotes the minimum value of* $\int_{K-K} |u| \cdot p(u)\, du$ *as* p *ranges over all elements of* $\mathcal{P}_n$ *non-negative on* $K - K$ *and satisfying* (2).

For example, take $m = 1$, $K = [-1/2, 1/2]$, so that $K - K = I = [-1, 1]$. We are led to the problem of computing $\lambda_n = \min \int_{-1}^{1} |t| \cdot p(t)\, dt$ over all non-negative polynomials of degree n such that $\int_{-1}^{1} p(t)\, dt = 1$. One can show, by the methods of Chapter 3 coupled with known estimates for the zeroes of Legendre polynomials, that $\lambda_n \leqslant c/(n + 1)$ (c = absolute constant), and hence deduce the *Jackson theorem*: *every* $f \in C[-1/2, 1/2]$ *is approximable by an element of* $\mathcal{P}_n$ *with an error not exceeding* $A\omega(f;\, (n + 1)^{-1})$ *where* A *is an absolute constant*.

The requirement, in the above analysis, that f vanish at the boundary (in the present example, therefore, at $t = \pm 1/2$) imposes here no restriction of generality if $n \geqslant 1$, since we can subtract from f a linear polynomial which interpolates it at $t = \pm 1/2$. In applying the method in more complex situations e.g. in higher dimensions, we must make a preliminary extension of the given function to some larger compact set on whose boundary it vanishes, so as to keep control of the m.o.c. of the extended function.

Exercises. a) Prove that there exists $p \in \mathcal{P}_n$ satisfying $\int_{-1}^{1} p(x)^2\, dx = 1$, $\int_{-1}^{1} x^2\, p(x)^2\, dx \leqslant c(n + 1)^{-2}$, where c is an absolute constant, and deduce from this the Jackson theorem (hint: use the mechanical quadrature formula of Chapter 3).

b) Paraphrase the above argument in terms of positive operators; deduce Jackson's theorem from 8.8.2.4 coupled with the preceding exercise.

8.8.2.6 Another variant. In the preceding paragraph, we saw how the proof of the fundamental degree-of-approximation theorem (that of Jackson) reduces to the search for a single polynomial satisfying a certain moment inequality. A variant method, which leads to a formally different problem, is also of interest. By means of a preliminary approximation (e.g. with a suitable broken-line function) it is easily seen that Jackson's theorem can be reduced to the case where f is of class Lip 1 on the basic interval $J = [-1/2, 1/2]$. It is also no real restriction to suppose $f(-1/2) = f(1/2) = 0$. Hence we may suppose f defined on all of R and vanishing outside J. Integrating by parts the identity $f = f * \delta$, where δ denotes the Dirac measure, gives $f = -(f' * h)$ where h is the "Heaviside function", $h(x) = (1 + \operatorname{sgn} x)/2$. Hence, for $p \in \mathcal{P}_n$,

$$f = -f' * (h - p) - f' * p$$

and so $q = -(f' * p) \in \mathcal{P}_n$ satisfies

$$|f(t) - q(t)| \leq \left| \int_{-\infty}^{\infty} f'(t-u)(h(u) - p(u))\, du \right|$$

whence

$$\max_{|t| \leq 1/2} |f(t) - q(t)| \leq \|f'\|_\infty \int_{-1}^{1} |h(u) - p(u)|\, du \ .$$

Therefore, in this variant, the Jackson theorem comes down to showing that h(x) <u>can be approximated on</u> $[-1, 1]$ <u>with an error</u> $O(n^{-1})$, <u>in the</u> L^1 <u>metric, by an element of</u> $\mathcal{P}_n$. This is a very typical situation: the problem of polynomial approximation of a certain function class (here the Lipschitz 1 class) in one metric leads to the problem of approximating <u>one fixed</u> ("kernel") <u>function</u> (characterizing that class) in the dual metric. In the present instance, the approximation of the kernel h(x) (or, what comes to the same after a change of variables, the problem of showing that on $[0, 1]$ the function $\varphi(x) = x^{-1/2}$ is approximable in L^1 by an element of $\mathcal{P}_n$ with error $O(1/n)$) could be carried out in various ways, but we shall not give details here as we prefer to follow the path of proving Jackson's theorem for <u>trigonometric</u> polynomials (in the next chapter) and then deduce the algebraic case by a change of variable. (The trigonometric case is very easy, involving essentially no computation, because of the leverage we get by application of Fourier analysis. Moreover, with the aid of Fourier methods we shall deduce a sharpened form of Jackson's theorem for

C(I), involving improved approximation near the endpoints (9.3.3.4).) Nevertheless, it is important to be in possession of direct methods for constructing (ordinary) polynomial approximations, especially to have flexibility in dealing with other situations (higher dimensions, approximation on peculiar sets, etc.). Moreover, the method of the present section can be adapted to constructing rational approximations, starting from Newman's result that $h(x)$ can be approximated in L^1 norm by rational functions of degree n with an error $O(e^{-\sqrt{n}})$. The rational case involves peculiar technical complications, however, and shall not be further discussed here.

Exercises. a) Consider the problem of approximating f on $[-1/2, 1/2]$ which satisfies $|f^{(m)}(x)| \leq 1$. What is the dual problem (in L^1) to which it leads?

b) Show that dist $(x^{-1/2}, \mathcal{P}_n)$ in $L^1[0, 1]$ is $O(1/n)$.

c) Exhibit functions h_1, h_2 such that $f_1 * h_1 + f_2 * h_2 = f$ holds for every $f \in C^1(R^2)$ vanishing outside the unit square. Use this to set up a "dual problem" to the problem of constructing polynomial approximations to functions satisfying $\|f_1\|_\infty + \|f_2\|_\infty \leq 1$ (here f_i denotes $\partial f/\partial x_i$). (Hint: take distributional Fourier transforms.)

Chapter 9. Approximation Theory in Homogeneous Banach Spaces

9.1 Background. At various points in the preceding lectures we have encountered (without, however, attempting a systematic discussion) particular Banach spaces of functions defined on spaces which admit a topological group of homeomorphisms on themselves (translations of R^n, rotations of spheres). We wish now to discuss this situation systematically in the context of R^n, and develop a framework within which we can quite economically formulate and prove "direct" and "inverse" theorems concerning approximation by convolution integrals, and by trigonometric and algebraic polynomials. The treatment we shall give is very "concentrated", as we avail ourselves simultaneously of two quite independent sources of unification. The first is the "Tauberian" viewpoint which we have expounded in Chapter 5 of our earlier book ($SHAPIRO_1$), endowed with considerable technical improvements from BOMAN & $SHAPIRO_2$. The second is the point of view of "homogeneous Banach spaces" (so christened in KATZNELSON) whereby approximation in a variety of norms is treated at one stroke (see 9.2, 9.3 for details). This very fruitful viewpoint seems to originate in the writings of Bochner; one can trace the beginnings already in his definition of almost-periodicity, and applications to approximation theory may be found in $BOCHNER_1$, Chapter 1. For the fullest exploitation of the underlying idea one requires the integration of vector-valued functions (a theory to which, once again, Bochner made pioneering contributions). We shall take for granted the elements of vector-valued integration (see e.g. HILLE & PHILLIPS).

Katznelson later developed the Bochner idea systematically, greatly extending its scope. Also, writings of Stechkin and some other Soviet analysts show a clear appreciation of the usefulness of these "homogeneous Banach spaces" as a unifying tool in approximation theory (cf. $STECHKIN_2$ and KUPTSOV).

We shall operate in spaces of functions defined on R^n, and handle simultaneously (i) all L^p norms, $1 \leq p \leq \infty$, as well as (ii) spaces both of periodic and non-periodic functions. So far as we know no treatment embodying feature (ii) has appeared in print, although such a treatment has the practical merit of eliminating a major source of duplication in approximation theory literature. Concerning (i), it should be remarked that the simultaneous treatment of all L^p norms within an axiomatic framework, with its attendant advantages, also encounters natural limits: in problems of more delicate character, subtle differences between the various L^p norms make themselves felt; on this point, see BOMAN & SHAPIRO$_{1,2}$. In these lectures we'll however, restrict ourselves to the coarser theory wherein these subtleties are not taken into account.

Finally we remark that some portion, at least, of the theory developed below can be carried out more generally, for spaces of functions defined on l.c.a. groups or on homogeneous spaces. However, here we shall consider as carrier space only R^n (and by implication the quotient space T^n thereof) in view of the great precision of the results attainable, and the importance of these cases in applications. Our mode of procedure is different in some details from that of Bochner or that of Katznelson in the works cited.

It is possible to give a relatively abstract, or a relatively concrete ("function space") definition of the class of spaces with which we shall deal. The abstract formulation is given in 9.2 and the more concrete formulation in 9.3.

9.2 Abstract homogeneous Banach spaces

9.2.1 Definition of an AHBS. Concerning functions and measures on R^n, and their Fourier transforms, we shall maintain the nomenclature of Chapter 7.

Let B denote a (for the time being, "abstract") Banach space, the norm in which shall be denoted by $\| \|_B$, or simply by $\| \|$ when there is no risk of ambiguity. We suppose that for each $u \in R^n$ there is defined a linear operator T_u on B, and that the map $u \to T_u$ is a homomorphism of the additive group R^n, i.e. T_0 is the identity, and $T_{u+v} = T_u T_v$. We shall impose three further assumptions, listed below as axioms (H1), (H2), (H3).

(H1) T_u *is an isometry of* B, *i.e.* $\|T_u f\| = \|f\|$, *all* $f \in B$.

(H2) *For each* $f \in B$, *the function* Φ_f *from* R^n *to* B *defined by*

(1) $$\Phi_f(u) = T_u f$$

is continuous.

Let now $\sigma \in M = M(R^n)$ (bounded complex measure on R^n), and $f \in B$. The element $f * \sigma$ of B is defined to be

(2) $$f * \sigma = \int \Phi_f(u)\, d\sigma(u) ,$$

the integral being understood as a Bochner-Stieltjes integral of the (continuous, bounded) B-valued function Φ_f. Observe that $f * (\sigma + \tau) = (f * \sigma) + (f * \tau)$, and a similar distributive law holds with respect to f. From basic properties of the integral we have also

(3) $$\|f * \sigma\|_B \leqslant \|\sigma\|_M \|f\|_B .$$

Observe that for every $k \in L^1 = L^1(R^n)$, since k is naturally identified as an element of M, we have a natural interpretation of $f * k$, for $f \in B$.

We now impose our final axiom:

(H3) <u>For every pair</u> $\sigma, \tau \in M$ <u>and every</u> $f \in B$,

(4) $$f * (\sigma * \tau) = (f * \sigma) * \tau .$$

Of course, in (4), the asterisk between σ and τ stands for convolution in M, whereas the remaining asterisks stand for the operation defined by (2).

<u>9.2.1.1 Definition</u>. An <u>abstract homogeneous Banach space</u> (abbreviated AHBS) <u>on</u> R^n is a Banach space B, together with a group $\{T_u\}$ of linear operators on B, such that (H1), (H2) and (H3) are satisfied.

<u>Remark</u>. Later we will want to consider a more particular situation, whereby B is a space of measurable functions on R^n and $\{T_u\}$ is the set of translations; however, we prefer to develop as much as we can of the necessary machinery in this abstract framework. The prototypical examples to keep in mind are $B = L^p(R^n)$ or $B = L^p(T^n)$, $1 \leqslant p < \infty$, with $(T_u f)(t) = f(t - u)$. In these cases (H2) is the well-known "continuity of translations" property. (H2) is not satisfied in $L^\infty(R^n)$, but it is in the closed subspace $\dot{C}(R^n)$ thereof, consisting of all bounded uniformly continuous functions on R^n. The validity of (H3) requires some discussion; it is a consequence of 9.3.2.3, together with the fact that the spaces just enumerated satisfy the axioms

for an HBS (see 9.3).

<u>9.2.2 Dilations of measures</u>. Let $C_o = C_o(R^n)$ denote the set of continuous functions on R^n which vanish at infinity. If $\sigma \in M$ and $a > 0$, the map $\varphi \to \int \varphi(au)\, d\sigma$ is a continuous linear functional on C_o, and hence the integral equals $\int \varphi(u)\, d\sigma_{(a)}(u)$ for some element $\sigma_{(a)}$ of M. This measure $\sigma_{(a)}$, we call *the a-dilation of* σ. For a function $k \in L^1(R^n)$, we write $k_{(a)}$ to denote the function defined by

$$k_{(a)}(t) = a^{-n}\, k(a^{-1}\, t) \ .$$

This is consistent with the preceding notation if k is interpreted in the natural way as an element of M.

We leave as exercises the proofs of the relations

(1) $$\|\sigma_{(a)}\|_M = \|\sigma\|_M \ , \quad a > 0$$

(2) $$(\sigma_{(a)})_{(b)} = \sigma_{(ab)} \ ,$$

(3) $$(\sigma * \tau)_{(a)} = \sigma_{(a)} * \tau_{(a)} \ .$$

Although the relation

(4) $$\int \varphi(u)\, d\sigma_{(a)}(u) = \int \varphi(au)\, d\sigma(u)$$

holds *a priori* only for $\varphi \in C_o$, it is easily shown that it remains valid also for all bounded continuous functions φ on R^n, in particular for the functions $\varphi = \varphi_x$: $t \to e^{ixt}$, for each $x \in \hat{R}^n$; therefore we have

(5) $$\hat{\sigma}_{(a)}(x) = \hat{\sigma}(ax) \ .$$

Moreover, we leave to the reader the verification that *(4) holds even when* φ *is a bounded continuous* B-*valued function*.

<u>9.2.2.1 Lemma</u>. *Let* B *be an AHBS, and suppose* $\sigma \in M$ *and* $\int d\sigma = 0$. *Then*

(1) $$\lim_{a \to 0} \|f * \sigma_{(a)}\| = 0 \ , \quad \textit{all } f \in B \ .$$

Proof. We have

$$f * \sigma_{(a)} = \int \Phi_f(u)\, d\sigma_{(a)}(u) = \int \Phi_f(au)\, d\sigma(u) = \int [\Phi_f(au) - \Phi_f(0)]\, d\sigma(u) .$$

Hence, denoting by $\tilde{\sigma}$ the total variation of σ,

$$(2) \qquad \|f * \sigma_{(a)}\| \leq \int \|T_{au}f - f\|\, d\tilde{\sigma}(u)$$

and since the (real-valued) continuous functions Ψ_a,

$$\Psi_a(u) \underset{\text{def.}}{=} \|T_{au}f - f\| ,$$

are uniformly bounded on R^n and satisfy $\lim_{a \to 0} \Psi_a(u) = 0$ for each $u \in R^n$, (1) follows from (2) by dominated convergence. ◇

An immediate corollary is

9.2.2.2. Lemma. Let B be an AHBS, and suppose $k \in L^1$ and $\int k\, dt = 1$. Then,

$$\lim_{a \to 0} \|f - (f * k_{(a)})\| = 0 .$$

Proof. Take for σ in 9.2.2.1 the measure $\delta - k\, dt$, δ = Dirac measure. ◇

9.2.3 Modulus of continuity

9.2.3.1 Definition. Let B be an AHBS and $f \in B$. The modulus of continuity of f is the function

$$\omega(a) = \omega(f;\, a) = \sup_{|u| \leq a} \|T_u f - f\| ; \quad a \geq 0 .$$

"Modulus of continuity" will be abbreviated m.o.c. Later we shall sometimes speak of the B m.o.c. in cases where a function f is presented in such a manner that there is possible ambiguity concerning the containing Banach space B we have in mind. When necessary, we shall employ such notations as $\omega(f;\, a)_B$, $\omega(f;\, a)_{C(T)}$, etc. to specify the m.o.c. in question.

9.2.3.2 Lemma. Let B be an AHBS, and $f \in B$. The m.c.c. $\omega(a) = \omega(f;\, a)$ is continuous on $[0, \infty)$, non-decreasing, subadditive, and bounded by $2\|f\|$; moreover, $\omega(0) = 0$, and for each $\lambda > 0$

(1) $$\omega(\lambda a) \leqslant (\lambda + 1)\, \omega(a) .$$

Proof. The continuity and boundedness, as well as that $\omega(0) = 0$, is evident from (H1) and (H2), and the monotonicity is a consequence of the definition of m.c.c. Now,

$$\|T_{u+v}f - f\| \leqslant \|T_{u+v}f - T_u f\| + \|T_u f - f\| = \|T_u(T_v f - f)\| + \|T_u f - f\|$$
$$= \|T_v f - f\| + \|T_u f - f\| .$$

Suppose now $a \geqslant 0$, $b \geqslant 0$. If u, v vary subject to the conditions $|u| \leqslant a$, $|v| \leqslant b$, the sum $u + v$ takes on all values in the ball $|t| \leqslant a + b$. Therefore, the above inequality implies $\omega(a + b) \leqslant \omega(a) + \omega(b)$, i.e. the subadditivity. By induction this implies $\omega(na) \leqslant n\omega(a)$, $n = 1, 2, \dots$. Hence, if n is the largest integer not exceeding λ, we have

$$\omega(\lambda a) \leqslant \omega((n + 1)a) \leqslant (n + 1)\, \omega(a) \leqslant (\lambda + 1)\, \omega(a) . \qquad \diamond$$

9.2.4 Lemma. *Let* B *be an AHBS*, $f \in B$, *and* $k \in L^1$, $\int k \, dt = 1$. *Then*

(1) $$\|f - (f * k_{(a)})\| \leqslant \omega(a) \int (1 + |u|)\, |k(u)|\, du$$

(observe that the integral is over R^n, *and* $|u|$ *denotes as usual the Euclidean norm of* u.)

Proof. Just as in the proof of 9.2.2.2.,

$$\|f - (f * k_{(a)})\| \leqslant \int \|f - T_{au}f\| \cdot |k(u)|\, du \leqslant \int \omega(f; a|u|)\, |k(u)|\, du \leqslant$$
$$\leqslant \omega(a) \int (1 + |u|)\, |k(u)|\, du . \qquad \diamond$$

9.2.4.1 Corollary. *Let* B *be an AHBS*, $f \in B$, $k \in L^1$, *and* $\int k \, dt = 1$. *Then*

$$\lim_{a \to 0} f * k_{(a)} = f .$$

Proof. From the preceding proof,

$$\|f - (f * k_{(a)})\| \leqslant \int \omega(f; a|u|)\, |k(u)|\, du$$

and the last integral tends to zero as $a \to 0$ by 9.2.3.2 and dominated convergence.

$\diamond$

9.3 Homogeneous Banach spaces of functions on R^n

9.3.1 Definition of a HBS. Throughout this section, B shall denote some Banach space, *the elements of which are (Lebesgue) measurable functions on* R^n, and T_u shall denote the operator of translation; that is, for $f \in B$,

$$(T_u f)(t) = f(t - u) \ , \quad u \in R^n \ .$$

Consider now the following axioms that a space B may satisfy:

(H'1) *For each* $u \in R^n$, T_u *is an isometry of* B *on itself*.

(H'2) "*Translation is continuous in* B", *i.e. for each* $f \in B$, *the map* $u \to T_u f$ *is a continuous* B-*valued function on* R^n.

(H'3) *The functions in* B *are uniformly locally integrable, i.e. there is some constant* α *such that*

$$\int_W |f(t - u)| \, dt \leq \alpha \, \|f\|_B \tag{1}$$

for every $f \in B$, $u \in R^n$; *here* W *denotes the cube* $\{0 \leq t_1 \leq 1; \ \ldots; \ 0 \leq t_n \leq 1\}$.

Observe that (1) implies that the analogous inequality holds with W replaced by an arbitrary compact set V, provided α is replaced by a suitable constant $\alpha(V)$.

9.3.1.1 Definition. A Banach space B consisting of Lebesgue measurable functions on R^n is a *homogeneous Banach space* (abbreviated HBS) if (H'1), (H'2) and (H'3) are satisfied.

Exercises. (Note: the notations introduced in these exercises will be used later without further commentary.) a) Prove that $\dot{C} = \dot{C}(R^n)$, the Banach space of bounded uniformly continuous functions on R^n, is a HBS.

b) Prove that $L^p(R^n)$ is a HBS, for $1 \leq p < \infty$.

c) Let $L^p(T^n)$ denote the Banach space of measurable functions on R^n which have period 2π in each variable, normed by

$$\|f\| = [(2\pi)^{-n} \int_0^{2\pi} \ldots \int_0^{2\pi} |f(t_1, \ldots, t_n)|^p \, dt_1 \ldots dt_n]^{1/p} \ .$$

Prove that, for $1 \leq p < \infty$, $L^p(T^n)$ is a HBS.

d) Let $S^p = S^p(R^1)$ denote the ("Stepanov") space of measurable functions on R^1 such that

$$\|f\|_{S^p} \underset{\text{def}}{=} \sup_{u \in R} \left(\int_u^{u+1} |f(t)|^p \, dt \right)^{1/p}$$

is finite. Prove that S^p is a Banach space, satisfying (H´1) and (H´3), but not (H´2). (Note: The study of closed subspaces of S^p which satisfy (H´2) is closely related to the theory of almost-periodic functions, cf. BOHR & FØLNER.)

e) Construct a Banach space of measurable functions on R^1 which satisfies (H´1) and (H´2) but not (H´3).

9.3.2 Relation of the concepts HBS and AHBS. In this section we shall prove that every HBS is an AHBS (relative, of course, to the group of translations). This is not obvious, because the axioms (H3) and (H´3) are very different in appearance.

We begin with

9.3.2.1 Lemma. *If f is any measurable function on R^n which is uniformly locally integrable, i.e. $\int_W |f(t - u)| \, dt \leq C$, where W is the unit cube of R^n, and C is independent of $u \in R^n$, and if $\sigma \in M(R^n)$, then for almost all $t \in R^n$*

(1)
$$g(t) = \int f(t - u) \, d\sigma(u)$$

exists as a Lebesgue-Stieltjes integral, and g is again uniformly locally integrable.

Proof. Clearly it is enough to give the proof for a positive measure σ. Now, for $s \in R^n$,

$$|g(t - s)| \leq \int |f(t - s - u)| \, d\sigma(u) .$$

$$\int_W |g(t-s)| dt \leq \int_W \left(\int |f(t-s-u)| d\sigma(u)\right) dt = \int \left(\int_W |f(t-s-u)| dt\right) d\sigma(u) \leq C \int d\sigma . \quad \diamond$$

In particular, for f in a HBS, (1) defines a locally integrable function g. Observe that (H´3) implies f is a tempered distribution, and one verifies that g is the convolution of f and σ in the sense of the theory of distributions. Now, the symbol $f * \sigma$ can also be given meaning, as an element of B in the sense of 9.2.1, i.e. as the vector-valued integral $\int (T_u f)\, d\sigma(u)$, and it is not *a priori* obvious that the two functions thus described are one and the same. We shall now prove that this is in fact the case.

9.3.2.2 Lemma. *Let* B *be a HBS, and* $f \in B$. *Suppose* $\sigma \in M$, *and define*

$$g(t) = \int f(t - u)\, d\sigma(u) \;,$$

$$h = \int (T_u f)\, d\sigma(u) \;.$$

The first integral denotes a Lebesgue-Stieltjes integral (existing a.e., by the preceding lemma), or equivalently, a distribution-theoretic convolution; the second, a B-valued integral. Then, $g(t) = h(t)$ *except on a set of Lebesgue measure zero.*

Proof. We shall, for simplicity, give the proof only for R^1. The general case involves more measure-theoretic sophistication, but can be carried out along the same lines. In the one-dimensional case, we can interpret σ as a function of bounded variation on R, and (by the definition of the Bochner-Stieltjes integral) we have, for suitably selected points

$$-\infty < u_{n,1} < u_{n,2} < \ldots < u_{n,N(n)} < \infty \;,$$

$$h = \lim_{n \to \infty} \sum_j (T_{u_{n,j}} f)(\sigma(u_{n,j+1}) - \sigma(u_{n,j}))$$

where the convergence is in the norm topology of B; here the summation is from $j = 1$ to $N(n)$. Let now φ be a continuous function vanishing outside a compact interval J. Since B-convergence implies convergence in the $L^1(J)$ topology,

$$\int h(t)\, \varphi(t)\, dt = \lim_{n \to \infty} \sum_j (\sigma(u_{n,j+1}) - \sigma(u_{n,j})) \int (T_{u_{n,j}} f)(t)\, \varphi(t)\, dt$$

$$= \lim_{n \to \infty} \sum_j (\sigma(u_{n,j+1}) - \sigma(u_{n,j})) \int f(t - u_{n,j})\, \varphi(t)\, dt \;.$$

Now, this relation is valid (in view of the definition of the integral) provided the $u_{n,j}$ satisfy the conditions $\lim_{n\to\infty} u_{n,1} = -\infty$, $\lim_{n\to\infty} u_{n,N(n)} = \infty$, and

$$\lim_{n\to\infty} \sup_{1 \leq j < N(n)} (u_{n,j+1} - u_{n,j}) = 0 .$$

We have therefore, writing $\psi(u) = \int f(t-u)\,\varphi(t)\,dt$ (observe that ψ is continuous and bounded on R)

$$\int h(t)\,\varphi(t)\,dt = \lim_{n\to\infty} \sum_j (\sigma(u_{n,j+1}) - \sigma(u_{n,j}))\,\psi(u_{n,j})$$

whenever $u_{n,j}$ satisfy the preceding conditions, and this implies

$$\int h(t)\,\varphi(t)\,dt = \int \psi(u)\,d\sigma(u) = \int\left(\int f(t-u)\,\varphi(t)\,dt\right) d\sigma(u) = \int g(t)\,\varphi(t)\,dt$$

by Fubini's theorem, and the desired conclusion follows. ◇

Therefore, whenever we have a homogeneous Banach space B, we may interpret the "convolution" of an element $f \in B$ with an element $\sigma \in M$, which we shall denote by $f * \sigma$, in either of the above two senses without ambiguity. This has important consequences. First of all, axiom (H3) is satisfied, i.e. 9.2.1 (4) holds, since both sides of the equality $f * (\sigma * \tau) = (f * \sigma) * \tau$ are meaningful (and equal!) when $*$ is interpreted throughout as convolution of tempered distributions. Secondly, the important inequality 9.2.1 (3) holds. Summarizing, we have

9.3.2.3 Theorem. Let B be an HBS. Then

(i) for $f \in B$, $\sigma \in M$ the Lebesgue-Stieltjes integral

(1) $$g(t) = \int f(t-u)\,d\sigma(u)$$

exists outside a t-set of Lebesgue measure zero, and the function g so defined is (apart from correction on a set of Lebesgue measure zero) an element of B, which satisfies

(2) $$\|g\| \leq \|\sigma\|_M \|f\| .$$

(ii) B *is an AHBS (and so, the results proved in 9.2 are applicable, with* $f * \sigma$ *interpreted as the function* g *in (1))*.

9.3.3 Jackson's theorem in HBS. As a simple but important first application, we observe from 9.2.4.1: *if* $k \in L^1$, $\int k\,dt = 1$, *then* $\lim_{a \to 0} f * k_{(a)} = f$, *for* f *in a HBS*. This is a rather general form of "approximate identity"; a quantitative version is 9.2.4 (1).

9.3.3.1 Theorem (Jackson's theorem for HBS). *Let* B *be any HBS on* R^1, *and suppose* $f \in B$ *has period* 2π. *Then, for each integer* $n \geq 1$, *there exists* $U \in \mathcal{T}_{n-1}$ *satisfying*

$$\|f - U\| \leq A\omega(f;\, 1/n) \tag{1}$$

where A *is an absolute constant*.

Proof. Let $k \in L^1(R)$ satisfy (i) $\int |t| \cdot |k(t)|\,dt < \infty$, (ii) $\int k\,dt = 1$ and (iii) $\hat{k}(x) = 0$, $|x| \geq 1$. (Such k obviously exist; we can choose any three times differentiable function K on $\hat{R}$ vanishing outside $[-1, 1]$ and satisfying $K(0) = 1$, and define k to be the inverse Fourier transform of K. Then $k(t) = O(|t|^{-3})$ at ∞, so (i) holds.) Define $U = f * k_{(1/n)}$. Now, the definition of HBS guarantees that $f \in L^1(T)$; also, the Fourier transform of $k_{(1/n)}$ vanishes for $|x| \geq n$, and from this fact it is easy to deduce $U \in \mathcal{T}_{n-1}$ (cf. SHAPIRO$_1$, pp. 48 ff.). (Observe that, by slight abuse of notation, we are here denoting by $\mathcal{T}_{n-1}$ the space of trig. polys. of degree at most $n - 1$ with *complex* coefficients; we have done this also on several occasions earlier.) Finally, 9.2.4 (1) with $a = 1/n$ yields (1), with

$$A = \int (1 + |u|)\ |k(u)|\ du\ .$$ ◇

Remarks. a) This method of proof obviously yields also the analogous result for HBS of functions on R^n which have period 2π in each variable. Moreover, it yields the corresponding result where we assume in place of periodicity that f has its spectrum in some discrete set Δ, and require that U shall be a trigonometric sum whose frequencies are chosen from Δ and have absolute value less than n. As this kind of generalization involves no new idea, we omit the details.

b) Choosing B to be C(T) we obtain of course the classical Jackson theorem for trigonometric polynomial approximation; we have only to observe that the m.o.c. in the HBS sense coincides here with the classical m.o.c. In like manner, one obtains the L^p versions of Jackson's theorem (cf. TIMAN).

c) The idea of obtaining an approximating trigonometric polynomial as a convolution in the above manner occurs in BOCHNER_2 (we regret that, owing to its peculiar context, this had escaped our notice when SHAPIRO_1 was written).

9.3.3.2 Concerning functions of higher smoothness, we have: *the Favard-Ahiezer-Krein theorem 7.5.2 is valid in HBS*, i.e. if $f \in B$ has period 2π and is the r-fold integral of some element of B (denoted by $f^{(r)}$) then for a suitable $U \in \mathcal{T}_{n-1}$ we have

$$\|f - U\| \leqslant \lambda_r \|f^{(r)}\| \, n^{-r}$$

where λ_r is as in 7.5.2. The proof we gave for 7.5.2 applies word for word. (In the same way, one can generalize many other results in earlier chapters to HBS; we shall not deal with all of these generalizations explicitly.)

9.3.3.3 *Algebraic polynomials*. Suppose now $f \in C(I)$ and ω denotes the (classical) m.o.c. of f. The function g: $g(t) = f(\cos t)$ is in C(T), $\omega(g, a) \leqslant \omega(f, a)$ and there exists $U \in \mathcal{T}_{n-1}$ satisfying $\|g - U\| \leqslant A\omega(g; 1/n)$. Moreover, since $g(-t) = g(t)$, one can assume that U is a "cosine polynomial" so that $U(t) = P(\cos t)$ for a suitable $P \in \mathcal{P}_{n-1}$. Therefore

$$\|f - P\|_{C(I)} = \|g - U\|_{C(T)} \leqslant A\omega(g; 1/n) \leqslant A\omega(f; 1/n)$$

for a suitable absolute constant A, which is Jackson's theorem for $\mathcal{P}_{n-1}$. It is of interest, as shown by Timan (cf. TIMAN) that a more careful analysis leads to a sharper estimate.

9.3.3.4 *Theorem (Timan)*. *Let* $f \in C(I)$. *For each* $n \geqslant 1$ *there exists* $P \in \mathcal{P}_{n-1}$ *such that*

(1) $$|f(x) - P(x)| \leqslant A'\omega(f; \delta_n(x)) \text{ , where}$$

(2) $$\delta_n(x) = n^{-2} + n^{-1}(1 - x^2)^{1/2}$$

and A′ is an absolute constant.

Remark. Since $\delta_n(x) \leq 2/n$ for $x \in I$, this result is clearly stronger than that in 9.3.3.3.

Proof. Let $g(t) = f(\cos t)$, and define $U \in \mathcal{T}_{n-1}$ by $U = g * k_{(1/n)}$, where k is as in the proof of 9.3.3.1, except that now we impose the extra restriction $\int t^2|k(t)|dt < \infty$. We have

$$(3) \qquad g(t) - U(t) = \int (g(t) - g(t - (u/n)))\, k(u)\, du .$$

Now, $|g(t) - g(t - a)| = |f(\cos t) - f(\cos(t - a))| \leq \omega(f; |\cos t - \cos(t - a)|)$, and since

$$|\cos t - \cos(t - a)| = |\cos t(1 - \cos a) - \sin t \sin a| \leq (a^2/2) + |a|\cdot|\sin t| ,$$

we obtain from (3), writing $\omega(\ )$ for $\omega(f;\ )$,

$$|g(t) - U(t)| \leq \int \omega((u^2/2n^2) + |u|\cdot|\sin t|/n)\, |k(u)|\, du = \int \omega(G(u))\, |k(u)|\, du$$

$$\leq \omega(s)\int (s^{-1}G(u) + 1)\, |k(u)|\, du = \omega(s)s^{-1} \int G(u)\, |k(u)|\, du + \omega(s) \int |k(u)|\, du$$

where $G(u) = (u^2/2n^2) + |u|\cdot|\sin t|/n$, and s is a positive quantity which is at our disposal. Let us choose

$$s = \int G(u)|k(u)|\, du = (2n^2)^{-1} \int u^2|k(u)|\, du + n^{-1}|\sin t| \int |u|\cdot|k(u)|\, du$$

$$\leq A_1(n^{-2} + n^{-1}|\sin t|)$$

where A_1 is an absolute constant. Hence, if $U(t) = P(\cos t)$, where $P \in \mathcal{P}_{n-1}$, we have finally

$$|f(\cos t) - P(\cos t)| \leq \omega(s)(1 + \int |k(u)|\, du) \leq A_2\, \omega(n^{-2} + n^{-1}|\sin t|)$$

and, putting $\cos t = x$, we have (1), where $A' = A_2$ is some absolute constant. ◇

9.3.4 Applications to Fourier analysis. Jackson′s theorem for HBS implies a result about the Fourier partial sums.

9.3.4.1 Definition. Let B be a HBS whose elements are measurable functions on R of period 2π, and $n \geq 0$ an integer. The Lebesgue constant Λ_n^B is the norm of the (Fourier partial sum) operator F_n which maps each function onto the partial sum of order n of its Fourier series,

$$\Lambda_n^B = \|F_n\|_{B \to B} = \sup_{\substack{f \in B \\ \|f\| = 1}} \|F_n f\|_B .$$

If B is $C(T)$, $L^\infty(T)$ or $L^1(T)$, Λ_n^B is the classical Lebesgue constant $\Lambda_n \sim (4/\pi^2) \log n$, discussed in 8.6. For $B = L^2(T)$, $\Lambda_n^B = 1$, and for $B = L^p(T)$, $1 < p < \infty$, $\Lambda_n^B \leq A_p < \infty$ for all n (cf. ZYGMUND).

9.3.4.2 Theorem. *Let* B *be a HBS whose elements are measurable functions on* R *of period* 2π. *Let* $f \in B$, *and denote by* s_n *the partial sum of order* n *of the Fourier series of* f. *Then*

$$\|f - s_{n-1}\| \leq (1 + \Lambda_n^B)\, A\, \omega(f, 1/n) \tag{1}$$

where A *is an absolute constant (the same as in 9.3.3.1 (1)).*

Proof. Let $U \in \mathcal{T}_{n-1}$ satisfy

$$\|f - U\| \leq A\omega(1/n) \tag{2}$$

where ω is the B m.o.c. of f; we know such U exists, by 9.3.3.1. Therefore, applying the operator F_{n-1},

$$\|s_{n-1} - U\| = \|F_{n-1}(f - U)\| \leq \Lambda_n^B\, A\omega(1/n) \tag{3}$$

and (2), (3) imply (1). ◇

9.3.4.3 Corollary. *Under the hypotheses of 9.3.4.2, if* $\lim_{n \to \infty} \Lambda_n^B\, \omega(f; 1/n) = 0$, *the Fourier series of* f *converges to* f, *in the* B *norm.*

In particular, for $B = C(T)$, we get the Dini-Lipschitz theorem: *if* $\omega(f; a) = o(|\log a|^{-1})$ *the Fourier series of* f *converges uniformly to* f.

Another criterion for norm-convergence of the Fourier series is in terms of the coefficients (cf. KATZNELSON, p. 52).

9.3.4.4 Theorem. *Let* B *be a HBS whose elements are measurable functions on* R *of* period 2π, and suppose the B norms of the characters $t \to e^{int}$ are bounded. Let $f \in B$, and denote the Fourier coefficients of f by $\{f\hat{}(n)\}_{n=-\infty}^{\infty}$. Suppose

$$(1) \qquad \lim_{\rho \to 1} \overline{\lim_{N \to \infty}} \sum_{N < |n| \leq \rho N} |f\hat{}(n)| = 0 .$$

(In particular, (1) holds if $f\hat{}(n) = O(1/n)$.) Then, the Fourier partial sums s_n of f converge to f in the B norm.

Proof. Choose $k_\rho \in L^1(R)$ so that its Fourier transform $K_\rho = \hat{k}_\rho$ equals one on $[-1, 1]$ and zero for $|x| \geq \rho > 1$, and is bounded by one in between (e.g. a "trapezoid function"). Then, * denoting convolution of tempered distributions on R,

$$(k_\rho)_{(1/N)} * f = s_N + r_N$$

where

$$r_N(t) = \sum_{N < |n| \leq \rho N} f\hat{}(n) K_\rho(n/N) e^{int} ;$$

hence

$$\|r_N\| \leq A \sum_{N < |n| \leq \rho N} |f\hat{}(n)| \underset{\text{def.}}{=} B_{N,\rho} .$$

(here A is an upper bound for the norms of the characters). We have, therefore

$$(2) \qquad \overline{\lim_{N \to \infty}} \|[(k_\rho)_{(1/N)} * f] - s_N\| \leq \overline{\lim_{N \to \infty}} B_{N,\rho} \underset{\text{def.}}{=} \psi(\rho) ,$$

where, by hypothesis, $\lim_{\rho \to 1} \psi(\rho) = 0$. Now, $(k_\rho)_{(1/N)} * f$ tends to f as $N \to \infty$ by 9.2.4.1. Therefore, from (2), $\overline{\lim_{N \to \infty}} \|f - s_N\| \leq \psi(\rho)$, and since $\rho > 1$ here is arbitrary, $\overline{\lim_{N \to \infty}} \|f - s_N\| = 0$. ◇

Remarks. a) The analogous theorem in several variables is also true, the proof being essentially the same.

b) This theorem is quite sharp; for example, if $\lim_{|n| \to \infty} \lambda_n = \infty$, one can find $f \in C(T)$ such that $|f\hat{}(n)| \leq \lambda_n(1 + |n|)^{-1}$ and yet the Fourier series for f diverges at a point (ZYGMUND, vol. 1, p. 304).

The m.o.c. corresponding to certain Banach norms other than the classical sup norm have interesting connections with other topics. Thus, the $L^2(T)$ (or $L^2(R)$) m.o.c. is related to absolute convergence of Fourier series (or Fourier transforms), while the $L^1(T)$ (or $L^1(R)$) m.o.c. is related to the rate of decay of Fourier coefficients (or Fourier transforms).

<u>9.3.4.5 Theorem (S. Bernstein)</u>. *If* $f \in L^2(T)$ *and the* $L^2(T)$ *m.o.c.* ω *of* f *satisfies*

$$\int_0^1 \omega(s)\, s^{-3/2}\, ds < \infty ,$$

(in particular, if $f \in C(T)$ *belongs to* Lip α *in the classical sense, for some* $\alpha > 1/2$*) then* $\sum |f^\wedge(n)| < \infty$.

<u>Proof</u>. Denote $f^\wedge(n)$ by c_n. Then, for $m \geq 0$, we have (with an easily understood notation)

$$S_m \underset{\text{def.}}{=} \sum_{|n|=2^m+1}^{2^{m+1}} |c_n| \leq 2^{(m+1)/2} \Big(\sum_{|n|=2^m+1}^{2^{m+1}} |c_n|^2 \Big)^{1/2} ,$$

and

$$\sum_{|n|=2^m+1}^{2^{m+1}} |c_n|^2 \leq \sum_{|n| \geq 2^m+1} |c_n|^2 = \| f - s_{2^m} \|^2_{L^2(T)} \leq A\omega(2^{-m})^2 ,$$

where A is an absolute constant, by 9.3.3.1, since in $L^2(T)$ norm the Fourier sum s_{2^m} gives at least as good approximation to f as any other trig. poly. of the same degree. Thus,

$$(1) \qquad \sum_{|n| \geq 2} |c_n| = \sum_{m=0}^{\infty} S_m \leq A^{1/2} \sum_{m=0}^{\infty} 2^{(m+1)/2}\, \omega(2^{-m})$$

and since

$$2^{m/2}(2 - \sqrt{2})\, \omega(2^{-m}) \leq \int_{2^{-m}}^{2^{-m+1}} \omega(t)\, t^{-3/2}\, dt$$

we obtain from (1)

$$\sum_{|n| \geq 2} |c_n| \leq A_1 \int_0^2 \omega(t)\, t^{-3/2}\, dt$$

where A_1 is an absolute constant, which implies the assertion. ◇

9.3.4.6 Corollary. If $f \in C(T)$ and the $C(T)$ m.o.c. ω of f satisfies $\int_0^1 \omega(t)^{1/2}\, t^{-1}\, dt < \infty$, (in particular, if $f \in \mathrm{Lip}\ \epsilon$ for some $\epsilon > 0$), and f is of bounded variation, then $\sum |f^\wedge(n)| < \infty$.

Proof. If f is of bounded variation, we have $\int_0^{2\pi} |f(t + a) - f(t)|\, dt \leq Ka$ for some constant K (see e.g. SHAPIRO$_1$, p. 40). Hence

$$(1/2\pi) \int_0^{2\pi} |f(t + a) - f(t)|^2\, dt \leq Ka\, \omega(f; a)_{C(T)}$$

which implies

$$\omega(f; a)_{L^2(T)} \leq K_1 [a\omega(f; a)_{C(T)}]^{1/2}$$

and 9.3.4.5 now implies the desired conclusion. ◇

9.3.4.7 The Bernstein theorem above is sharp, e.g. in the sense that there exists a function $f \in C(T)$ belonging to Lip 1/2, such that $\sum |f^\wedge(n)| = \infty$.

To prove this, we use the fact that there exists a sequence $\{a_n\}_{n=0}^{\infty}$, where each a_n is 1 or -1, such that

$$(1) \qquad \left|\sum_{k=0}^{n} a_k e^{ikt}\right| \leq A(n + 1)^{1/2}, \quad n = 0, 1, \ldots$$

uniformly in t; here A is an absolute constant. (This was proved by me in 1950, and announced much later in SHAPIRO$_9$. Cf. KATZNELSON, p. 33, KAHANE & SALEM pp. 134 ff., KAHANE; other unimodular sequences would work equally well, cf. ZYGMUND, vol. 1, p. 197 ff., however the example cited is the most elementary.)

Consider now the series $\sum_{k=0}^{\infty} (a_k/(k + 1))\, e^{ikt}$. It follows easily by Abel partial summation, using (1), that the series converges uniformly, hence defines a function $f \in C(T)$. Moreover, denoting by s_n the partial sum of order n of this series, another partial summation shows $\|f - s_n\|_{C(T)} = O(n^{-1/2})$; hence by an "inverse theorem" of approximation, which we shall prove later (cf. also SHAPIRO$_1$, p. 69), $f \in \mathrm{Lip}\ 1/2$. Since $|f^\wedge(n)| = 1/(n+1)$, we have the desired counter-example.

An interesting use of the $L^1(R)$ m.o.c. is

9.3.4.8 Theorem. For $f \in L^1(R)$,

$$|\hat{f}(x)| \leqslant (1/2)\ \omega(f, \pi\ |x|^{-1})_{L^1(R)} .$$

Proof. It is clearly sufficient to deal with $x > 0$. Now, $\hat{f}(x) = \int_{-\infty}^{\infty} f(t)\ e^{-itx}\ dt$, hence

$$\hat{f}(x) = -\int_{-\infty}^{\infty} f(t + (\pi/x))\ e^{-itx}\ dt ,$$

and adding,

$$2\hat{f}(x) = \int_{-\infty}^{\infty} (f(t) - f(t + (\pi/x)))\ e^{-itx}\ dt$$

whence

$$|\hat{f}(x)| \leqslant (1/2) \int |f(t) - f(t + (\pi/x))|\ dt \leqslant (1/2)\ \omega(f, \pi/x)_{L^1(R)} . \quad \diamond$$

9.3.4.9 Exercises. a) Prove that $A(\hat{R})$ is a HBS.

b) Check which of the axioms for HBS are satisfied by each of the following spaces of periodic functions: Lip α (with norm $\|f\| = \sup_{t_1,t_2} (|f(t_1) - f(t_2)|/|t_1 - t_2|^\alpha)$); its closed subspace lip α consisting of f such that $f(t + a) - f(t) = o(|a|^\alpha)$ uniformly in t; the space V_p of functions whose Fourier coefficients are summable to the power p ($1 \leqslant p \leqslant 2$), with the obvious norm based on the coefficients. (Note: for certain fairly well known spaces, the verification of axiom (H´2) presents great difficulties, e.g. it is not known whether, in the "multiplier" space M_p of functions f on the circle such that $fg \in V_p$ whenever $g \in V_p$ (where $\|f\|_{M_p}$ is defined as the norm of the operator $g \to fg$ from V_p to V_p), translations are continuous. Here M_p is a Banach algebra, and the question at hand is a stumbling block to the study of its maximal ideal space.)

c) Same question as b) for r times differentiable functions, with usual Banach topology.

d) Prove that, if B is a space of periodic functions which satisfies (H´1) and (H´3), and contains all characters, then (H´2) holds if and only if the trigonometric polynomials are dense in B.

e) Prove that if f belongs to an AHBS and its m.o.c. is $o(a)$, then $T_u f = f$ for all $u \in R^n$; in particular, if f belongs to a HBS and $\omega(f, a) = o(a)$, f is constant.

f) Prove the remark "Observe ... " following the formulation of (H´3).

g) Prove 9.3.2.2 in R^n.

h) Can you prove anything similar to 9.3.3.4 in the $L^p(I)$ metric?

i) Prove 9.3.4.5 without using Jackson´s theorem for L^2.

j) Carry out the omitted estimations in 9.3.4.7. Also, prove that $f \in \text{Lip } 1/2$ without quoting the "inverse theorem".

k) Prove that the $L^1(T)$ m.o.c. of the function $f(t) = e^{i/|t|}$, $|t| \leq \pi$ is $O(a^{1/2})$.

l) Prove the analog of 9.3.4.8 for Fourier coefficients. What estimate does this give for the coefficients of the function defined in the preceding exercise? Is this a good estimate?

m) Let W denote any compact subset of $L^1(T)$. Prove that

$$\lim_{|n| \to \infty} \sup_{f \in W} |f^{\wedge}(n)| = 0 ,$$

(i) using the preceding exercise, and (ii) not using it, but rather 8.2.1.(Hint for (ii): $|f^{\wedge}(n)| = |(2\pi)^{-1} \int_0^{2\pi} f(t)\, e^{-int}\, dt| = |(2\pi)^{-1} \int_0^{2\pi} (f(t) - U(t))\, e^{-int}\, dt| \leq \|f - U\|_{L^1(T)}$, for every $U \in \mathcal{T}_{n-1}$.)

n) Prove that if $f \in L^{\infty}(T)$ and $f^{\wedge}(n) = O(1/n)$, the partial sums of the Fourier series of f are uniformly bounded. (Hint: convolve f with a suitable "trapezoid" kernel.)

o) Prove the following form of "Fourier inversion": <u>Suppose</u> $K \in L^1(\hat{R})$ <u>and</u> $k(t) = (1/2\pi) \int K(x)\, e^{itx}\, dx$ <u>belongs to</u> $L^1(R)$ <u>and</u> $\int k\, dt = 1$. <u>Then, for all</u> $f \in L^1(R)$ <u>we have</u>: <u>if</u> $f_\varepsilon(t)$ <u>denotes</u> $(1/2\pi) \int \hat{f}(x)\, K(\varepsilon x)\, e^{itx}\, dt$, $\lim_{\varepsilon \to 0} \|f_\varepsilon - f\|_{L^1(R)} = 0$.

p) Deduce from o) that if $f \in L^1(R)$ and $\hat{f} \in L^1(\hat{R})$, then

$$f(t) = (2\pi)^{-1} \int \hat{f}(x)\, e^{itx}\, dx \text{ holds a.e.}$$

q) Prove that if B is a HBS on R, containing the function f of period 2π, the Fejér sums σ_n formed from f ($\sigma_n = (s_0 + \dots + s_n)/(n+1)$) converge to f in the B-norm. (Hint: establish the formula $\sigma_{n-1}(t) = (f * k_{(1/n)})(t)$ where $k(t) = (2/\pi t^2)\sin^2(t/2)$.)

r) Prove the analogous proposition for the Abel means $f(r, t)$, based on the formula $f(e^{-a}, t) = (f * k_{(a)})(t)$ with $k(t) = (1/\pi)(1 + t^2)^{-1}$. (<u>Note</u>: the proofs suggested in q) and r) are not the usual ones; the study of summability of Fourier series by means of convolutions on the infinite line is often useful, especially in multidimensional problems.)

9.4 Comparison theorems

The material in this section is based on BOMAN & SHAPIRO$_2$.

9.4.1 The notion of σ-modulus

<u>9.4.1.1 Definition</u>. Let B be any AHBS and $f \in B$. For $\sigma \in M = M(R^n)$, <u>the σ-modulus of</u> f, denoted by $\omega_\sigma(f; a)$ or $\omega_{\sigma,B}(f; a)$ is the function on $[0, \infty)$ defined by

(1) $$\omega_\sigma(f; a) = \sup_{0 < b \leq a} \|f * \sigma_{(b)}\| .$$

(For the meaning of $\sigma_{(b)}$, see 9.2.2.)

Clearly, for fixed $f \in B$, $\sigma \in M$ the σ-modulus is a non-decreasing function of a. Moreover, it satisfies the following relations, whose verification we leave to the reader.

(2) $$\omega_{\sigma * \tau}(f; a) \leq \|\sigma\|_M\, \omega_\tau(f; a)$$

(3) $$\omega_{\sigma+\tau}(f; a) \leq \omega_\sigma(f; a) + \omega_\tau(f; a) .$$

9.4.2 A special partition of unity.

9.4.2.1 Lemma. There exists a function Φ on $\hat{R}^n$ which is infinitely differentiable at all points, and such that further

(i) $\Phi(x) > 0$ for $1/2 < |x| < 2$, and Φ vanishes elsewhere,

(ii) $\sum_{j=-\infty}^{\infty} \Phi(2^j x) = 1$ if $x \neq 0$.

Proof. Let $h(a)$ denote a function defined for $0 \leqslant a < \infty$ and equal to one for $0 \leqslant a \leqslant 1$, to zero for $a \geqslant 2$, strictly decreasing on $[1,2]$, and infinitely differentiable. Then $\Phi(x) = h(|x|) - h(2|x|)$ satisfies the requirements. Observe that in the series (ii) at most two terms are different from zero, for each $x \neq 0$. $\diamond$

Φ is the Fourier transform of a certain function φ which is infinitely differentiable, and φ and all its partial derivatives tend to zero at infinity more rapidly than any negative power of $|t|$.

We shall, throughout, write φ_j to denote $\varphi_{(2^j)}$, i.e.

(1) $$\varphi_j(t) = \varphi_{(2^j)}(t) = 2^{-nj}\,\varphi(2^{-j}t) , \quad j = 0, \pm 1, \ldots$$

Observe that

(2) $$\hat{\varphi}_j(x) = \Phi(2^j x) , \quad j = 0, \pm 1, \ldots$$

We shall also require the relation, for positive integral r,

(3) $$\sum_{j=-r}^{r} \Phi(2^j x) = 1 \quad \text{for} \quad 2^{-r} \leqslant |x| \leqslant 2^r ,$$

which follows from (ii), since for x in this range, and $|j| > r$, $\Phi(2^j x) = 0$.

In the analysis which follows, we shall always suppose the dimension n and the choice of a particular Φ with the properties enumerated in the lemma to have been fixed, and treat as "constants" numbers which depend only on n and Φ.

9.4.3 "Tauberian condition". A continuous function F on $\hat{R}^n$ is said to satisfy the Tauberian condition if, for every x with $|x| = 1$, there exists $c \geqslant 0$ such that

$F(cx) \neq 0$, in other words, if F takes a non-zero value on every closed half-ray. (If, in particular, $F(0) \neq 0$, F satisfies the Tauberian condition trivially.)

9.4.3.1 Lemma. *If F is continuous on $\hat{R}^n$ and satisfies the Tauberian condition, and $\delta > 0$ is arbitrary, there exist positive numbers $d_1 < d_2 \dots < d_r$ such that*

$$\sum_{j=1}^{r} |F(d_j x)| > 0 , \quad \delta \leqslant |x| \leqslant 1/\delta .$$

Proof. Let S denote $\{x : |x| = 1\}$ and F_c, for $c > 0$, the function on S defined by $F_c(x) = F(cx)$. The hypotheses imply $\{F_c\}_{c>0}$ have no common zero on S, i.e. the closed subsets E_c of S defined by $E_c = \{x : F_c(x) = 0\}$ have an empty intersection. Therefore, since S is compact, there is some finite subset $E_{c_1}, \dots, E_{c_m}$ of the E_c whose intersection is empty. This means that $G(x) = \sum_{i=1}^{m} |F(c_i x)|$ is positive on S, and hence by continuity remains positive in the spherical shell $b \leqslant |x| \leqslant 1/b$ if b is chosen sufficiently close to 1. Therefore, if k is chosen large enough, $\sum_{j=-k}^{k} G(b^j x)$ is positive for $\delta \leqslant |x| \leqslant 1/\delta$, which implies the assertion in the lemma. ◇

9.4.4 Comparison of two moduli.

9.4.4.1 Lemma. *Let B be any AHBS, $f \in B$ and $\sigma, \tau \in M(R^n)$. Suppose moreover $\hat{\sigma}$ satisfies the Tauberian condition, and $\hat{\tau}$ vanishes in neighborhoods of 0 and ∞. Then*

(1)
$$\omega_\tau(f; a) \leqslant A_1\, \omega_\sigma(f; A_2\, a)$$

where A_1, A_2 are positive constants depending only on σ, τ.

Proof. By 9.4.3.1 there exist numbers $d_j > 0$ such that $\sum_{j=1}^{r} |\hat{\sigma}(d_j x)|$ is positive on the support of $\hat{\tau}$. Hence, by the basic ideal theory of $M(R^n)$, τ belongs to the ideal in $M(R^n)$ generated by $\sigma_{(d_1)}, \dots, \sigma_{(d_r)}$, i.e.

$$\tau = \sum_{j=1}^{r} \rho_j * \sigma_{(d_j)} , \quad \text{where} \quad \rho_j \in M(R^n) .$$

Hence, for $a > 0$

$$\tau_{(a)} = \sum_{j=1}^{r} (\rho_j)_{(a)} * \sigma_{(ad_j)} .$$

Therefore, for $f \in B$,

$$\|f * \tau_{(a)}\| \leq \sum_{j=1}^{r} \|\rho_j\|_M \|f * \sigma_{(ad_j)}\| \leq \sum_{j=1}^{r} \|\rho_j\|_M \omega_\sigma(f; ad_j) \leq A_1 \omega_\sigma(f; A_2 a)$$

and now (1) follows. ◇

We work from now on in the slightly lesser generality of HBS, because we will require that the elements of the space under consideration be tempered distributions.

<u>9.4.4.2 Lemma</u>. <u>Let</u> B <u>be any HBS</u>, $f \in B$ <u>(in particular</u>, f <u>is locally integrable, and moreover is a tempered distribution on</u> R^n), <u>and suppose the support of</u> $\hat{f}$ <u>does not contain the origin</u>. <u>For any</u> $b > 0$ <u>we have</u>

(1)
$$\|f\| \leq \sum_{j=-\infty}^{\infty} \|f * \varphi_{(2^j b)}\| .$$

<u>Proof</u>. Suppose first $\hat{f}$ has compact support. Then, for a suitably large integer N (depending on b and supp $\hat{f}$)

$$f = \sum_{j=-N}^{N} f * \varphi_{(2^j b)}$$

since the Fourier transform of $\sum_{j=-N}^{N} \varphi_{(2^j b)}$, namely $\sum_{j=-N}^{N} \Phi(2^j bx)$, equals 1 on a neighborhood of supp $\hat{f}$. Hence

$$\|f\| \leq \sum_{j=-N}^{N} \|f * \varphi_{(2^j b)}\| \leq \sum_{j=-\infty}^{\infty} \|f * \varphi_{(2^j b)}\| .$$

In the general case, choose $k(t) \geq 0$ such that $\int k \, dt = 1$ and $\hat{k}$ has compact support; then $f * k_{(a)}$ has a (distributional) Fourier transform whose support is compact and does not contain the origin, so we may apply the above to it and obtain

$$\|f * k_{(a)}\| \leq \sum_{j=-\infty}^{\infty} \|f * k_{(a)} * \varphi_{(2^j b)}\| \leq \sum_{j=-\infty}^{\infty} \|f * \varphi_{(2^j b)}\| ,$$

and now, letting $a \to 0$ and observing that $\|f * k_{(a)}\| \to \|f\|$ (see 9.2.4.1), we obtain (1). ◇

For later purposes, it is important to observe that the restriction on the support of $\hat{f}$ is not really essential; indeed:

_If_ B _does not contain the function identically equal to_ 1, _then_ (1) _holds for all_ $f \in B$. _Moreover, in any case,_ (1) _holds with_ f _replaced by_ $g * \tau$ _provided_ $g \in B$, $\tau \in M$ _and_ $\hat{\tau}(0) = 0$.

To see this, observe that (1) holds trivially unless

$$\sum_{j=-\infty}^{\infty} \|f * \varphi_{(2^j b)}\| < \infty . \tag{2}$$

We may therefore suppose that (2) holds. Hence

$$\sum_{j=-\infty}^{\infty} f * \varphi_{(2^j b)}$$

converges (in B) to an element which we denote by h. If ψ is any element of $L^1(R^n)$ whose Fourier transform vanishes in neighborhoods of 0 and ∞, then

$$h * \psi = \sum_{j=-\infty}^{\infty} (f * \psi) * \varphi_{(2^j b)} = f * \psi ;$$

(cf. the proof of Lemma 9.4.4.2) hence the Fourier transform of $f - h$ is supported at the origin, so $f - h$ is a polynomial. Since $f - h$, being an element of B, is uniformly locally integrable, it reduces to a constant. Hence, if $\underline{1} \notin B$, $f = \sum f * \varphi_{(2^j b)}$ and (1) holds. If $\underline{1} \in B$, we have in any case, for each $g \in B$ and some constant $\lambda = \lambda(g)$,

$$g = \lambda \underline{1} + \sum_{j=-\infty}^{\infty} g * \varphi_{(2^j b)} .$$

Thus, if $\hat{\tau}(0) = 0$,

$$g * \tau = \sum_{j=-\infty}^{\infty} (g * \tau) * \varphi_{(2^j b)}$$

yielding again (1), with $g * \tau$ in place of f. ◇

We come now to the most important lemma.

9.4.4.3 Lemma. Let B be any HBS, $f \in B$ and $\sigma, \tau \in M(R^n)$. Suppose moreover $\hat{\sigma}$ satisfies the Tauberian condition, and $\hat{\tau}$ vanishes in a neighborhood of 0. Then

$$(1) \qquad \omega_\tau(f; a) \leq A \int_0^{Ba} \omega_\sigma(f; s)\, ds/s$$

where A, B depend only on σ, τ.

Remark. The difference, small but crucial, between this lemma and 9.4.4.1 is that now $\hat{\tau}$ is not required to have compact support; observe that (1) is, in general, weaker than 9.4.4.1 (1).

Proof of lemma. For any $a > 0$, $f * \tau_{(a)}$ has a Fourier transform whose support does not contain the origin. Hence, by the preceding lemma, for any $b > 0$

$$(2) \qquad \|f * \tau_{(a)}\| \leq \sum_{j=-\infty}^{\infty} \|f * \tau_{(a)} * \varphi_{(2^j b)}\| .$$

Suppose now $\hat{\tau}(x) = 0$ for $|x| \leq c$, then $\hat{\tau}_{(a)}(x) = 0$ for $|x| \leq c/a$. Since the Fourier transform of $\varphi_{(2^j b)}$ vanishes outside $\{2^{-j-1} b^{-1} \leq |x| \leq 2^{-j+1} b^{-1}\}$ we see that if we choose

$$(3) \qquad b = 2a/c ,$$

$\tau_{(a)} * \varphi_{(2^j b)}$ is zero for $j \geq 0$. Thus, defining b by (3), the summands in (2) with $j \geq 0$ are zero, and we get

$$(4) \qquad \|f * \tau_{(a)}\| \leq \sum_{j=-\infty}^{-1} \|f * \tau_{(a)} * \varphi_{(2^j b)}\| \leq \|\tau\|_M \sum_{j=-\infty}^{-1} \|f * \varphi_{(2^j b)}\| .$$

Now, by Lemma 9.4.4.1, taking φdt in place of the measure τ in that lemma, we have

$$\|f * \varphi_{(2^j b)}\| \leq A_1\, \omega_\sigma(f; A_2\, 2^j b)$$

where A_1, A_2 depend only on σ and φ (hence effectively only on σ, φ being supposed fixed once for all). Thus, from (4)

$$\|f * \tau_{(a)}\| \leq A_1\, \|\tau\|_M \sum_{j=1}^{\infty} \omega_\sigma(f; A_2\, 2^{-j}\, b)$$

and observing the inequality

$$(\log 2)\, \omega_\sigma(f; A_2\, 2^{-j}\, b) \leq \int_{2^{-j}}^{2^{-j+1}} \omega_\sigma(f; A_2\, bs)\, ds/s = \int_{2^{-j}A_2 b}^{2^{-j+1}A_2 b} \omega_\sigma(f; u)\, du/u \;,$$

summing on j, and substituting the value of b from (3), we obtain (1), with $A = (\log 2)^{-1}\, A_1 \|\tau\|_M$ and $B = 2A_2/c$. ◇

From this result we deduce easily our first "comparison theorem" for two moduli.

<u>9.4.4.4. Theorem</u>. <u>Let</u> B <u>be any</u> HBS, $f \in B$, <u>and</u> $\sigma, \tau \in M(R^n)$. <u>Suppose moreover</u> $\hat{\sigma}$ <u>satisfies the Tauberian condition, and</u> $\hat{\sigma}$ <u>divides</u> $\hat{\tau}$ <u>at</u> $x = 0$ (see 7.3.11 for the terminology). <u>Then</u>

(1) $$\omega_\tau(f; a) \leq A \int_0^{Ba} \omega_\sigma(f; u)\, du/u$$

<u>where</u> A, B <u>depend only on</u> σ, τ.

<u>Proof</u>. For some $\mu \in M$, the measure $\rho = \tau - \mu * \sigma$ has a Fourier transform which vanishes in a neighborhood of 0, hence by 9.4.4.3

(2) $$\omega_\rho(f; a) \leq A_0 \int_0^{B_0 a} \omega_\sigma(f; u)\, du/u$$

where A_0, B_0 depend only on σ and ρ, i.e. only on σ and τ. Now, by 9.4.1.1 (2) and (3),

$$\omega_\tau(f; a) \leqslant \|\mu\|_M \cdot \omega_\sigma(f; a) + \omega_\rho(f; a) \tag{3}$$

and

$$\omega_\sigma(f; a) \leqslant \int_a^{ea} \omega_\sigma(f; u)\, du/u . \tag{4}$$

Substituting into (3) the estimates for ω_σ and ω_ρ from (4), (2) respectively, gives (1). ◇

By a slight generalization of the arguments leading up to 9.4.4.4 one can prove the more general proposition: _if_ $\tau, \sigma_1, \ldots, \sigma_r \in M(R^n)$, $\sum_{j=1}^r |\hat{\sigma}_j(x)|$ _satisfies the Tauberian condition, and_ $\hat{\tau}$ _belongs locally_ (_at_ $x = 0$) _to the ideal generated by_ $\hat{\sigma}_1, \ldots, \hat{\sigma}_r$, _then_

$$\omega_\tau(f; a) \leqslant A \int_0^{Ba} \Big(\sum_{j=1}^r \omega_{\sigma_j}(f; u) \Big)\, du/u \tag{5}$$

where A, B _depend only on_ $\tau, \sigma_1, \ldots, \sigma_r$. The details are left to the reader.

The point of Theorem 9.4.4.4 is that only a _local_ divisibility condition is imposed; in case $\hat{\sigma}$ divides $\hat{\tau}$ (globally), we have in place of (1), and without assuming $\hat{\sigma}$ satisfies the Tauberian condition,

$$\omega_\tau(f; a) \leqslant A_1\, \omega_\sigma(f; B_1 a) \tag{6}$$

with suitable constants A_1, B_1. This is a consequence of the relatively trivial inequality 9.4.1.1 (2). What is striking is that (1) is nearly as strong as (6), and fully as useful for many applications. We shall discuss later other hypotheses under which (6) holds. First, however, we shall deduce from 9.4.4.4 another comparison theorem, whereby τ is slightly more restricted but in exchange the local divisibility hypothesis can be dropped.

9.4.4.5 Theorem. _Suppose_ $\sigma, \tau \in M(R^n)$ _and_ $\hat{\sigma}$ _satisfies the Tauberian condition. Let_ P _be a positive-homogeneous function on_ $\hat{R}^n$ _of degree_ r (_i.e._ $P(cx) = c^r P(x)$ _for_ $c > 0$), _where_ r _is a positive real number_, $P \in C^\infty(R^n \setminus \{0\})$, _and suppose for some_ $\theta \in M(R^n)$ _we have_ $\hat{\tau}(x) = \hat{\theta}(x)\, P(x)$ _for all_ x _in some neighborhood of the origin. Then, for_ $f \in B$,

(1) $$\omega_\tau(f;\,a) \leqslant A \int_0^\infty [\min(1,\,a/u)]^r\,\omega_\sigma(f;\,u)\,du/u$$

<u>where</u> A <u>depends only on</u> σ, τ.

<u>Proof</u>. We may suppose that $\hat{\theta}$ has compact support, since this can be achieved upon multiplication by an element of $B(\hat{R}^n)$ which has compact support and equals 1 in a neighborhood of 0. We can then write $\tau = \mu + \nu$, where $\mu, \nu \in M(R^n)$, and

(2) $$\hat{\mu}(x) = P(x)\,\hat{\theta}(x)\ , \quad \text{all } x \in \hat{R}^n$$

(3) $$\hat{\nu}(x) \text{ vanishes in a neighborhood of } 0.$$

For, $P\hat{\theta}$ belongs to $B(\hat{R}^n)$ at each point of $\hat{R}^n$, including infinity (cf. the discussion in 7.3.11), hence $P\hat{\theta}$ is an element of $B(\hat{R}^n)$; denoting it by $\hat{\mu}$, $\hat{\nu} \underset{\text{def.}}{=} \hat{\tau} - \hat{\mu}$ clearly satisfies (3). Now,

(4) $$\omega_\tau(f;\,a) \leqslant \omega_\mu(f;\,a) + \omega_\nu(f;\,a)$$

and, because of (3) and 9.4.4.3

(5) $$\omega_\nu(f;\,a) \leqslant A_1 \int_0^{B_1 a} \omega_\sigma(f;\,u)\,du/u$$

for suitable constants A_1, B_1. If A is chosen large enough, the right side of (1) exceeds that of (5). Hence, in view of (4), it suffices to prove (1) with ω_μ in place of ω_τ. We have, for $c > 0$, by the generalization of Lemma 9.4.4.2,

(6) $$\|f * \mu_{(a)}\| \leqslant \sum_{j=-\infty}^{\infty} \|f * \mu_{(a)} * \varphi_{(2^j ca)}\|\ .$$

Since $\hat{\mu}$ has compact support and $\hat{\varphi}(x) = 0$ for $|x| \leqslant 1/2$, we can choose c (independent of a) such that $\mu_{(a)} * \varphi_{(2^j ca)} = 0$ for $j \leqslant -1$; <u>making this choice of</u> c, <u>we can ignore those summands in (6) with</u> $j < 0$. Now,

$$(\mu_{(a)} * \varphi_{(2^j ca)})^\wedge\,(x) = \hat{\mu}(ax)\,\hat{\varphi}(2^j cax) = P(ax)\,\hat{\theta}(ax)\,\hat{\varphi}(2^j cax)$$

$$= P(2^j cax)\,\hat{\varphi}(2^j cax)\,\hat{\theta}(ax)(2^j c)^{-r} = (2^j c)^{-r}\,G(2^j cax)\,\hat{\theta}(ax)$$

where $G = P\hat{\varphi}$ is of class $C^\infty(\hat{R}^n)$ and supported in $\{1/2 \leqslant |x| \leqslant 2\}$. Writing $G = \hat{g}$, with $g \in L^1(R^n)$, we have therefore

$$\|f * \mu_{(a)}\| \leqslant \sum_{j=0}^{\infty} \|f * \mu_{(a)} * \varphi_{(2^j ca)}\| = \sum_{j=0}^{\infty} (2^j c)^{-r} \|f * g_{(2^j ca)} * \theta_{(a)}\|$$

$$\leqslant c^{-r} \|\theta\|_M \sum_{j=0}^{\infty} 2^{-jr} \|f * g_{(2^j ca)}\| .$$

Moreover, by 9.4.4.1 (taking $g\, dt$ in place of the τ in that lemma), for any $b > 0$

$$\|f * g_{(b)}\| \leqslant A_2\, \omega_\sigma(f;\, A_3 b) ,$$

where A_2, A_3 (as A_4, A_5, ... below) depend only on σ, τ. Hence

$$\|f * \mu_{(a)}\| \leqslant c^{-r} \|\theta\|_M A_2 \sum_{j=0}^{\infty} 2^{-jr}\, \omega_\sigma(f;\, A_3\, 2^j ca) = A_4 \sum_{j=0}^{\infty} 2^{-jr}\, \omega_\sigma(f;\, A_5\, 2^j a) .$$

Because of the inequality

$$\int_{2^j}^{2^{j+1}} u^{-r}\, \omega_\sigma(f;\, A_5\, au)\, du/u \geqslant 2^{-r(j+1)}\, \omega_\sigma(f;\, A_5\, 2^j\, a) \log 2 ,$$

we have

$$\|f * \mu_{(a)}\| \leqslant A_6 \int_1^{\infty} u^{-r}\, \omega_\sigma(f;\, A_5\, au)\, du/u = A_7\, a^r \int_{A_5 a}^{\infty} u^{-r}\, \omega_\sigma(f;\, u)\, du/u$$

and the right side of (1) exceeds this last expression if A is chosen large enough. This yields the needed estimate for $\omega_\mu(f;\, a)$, and in view of (4), (5) the proof is complete. ◇

<u>9.4.4.6 Refinements of the preceding theorems</u>. Recently Jan Boman has shown that if, in Theorem 9.4.4.4, σ is subjected to the additional requirement to belong to a certain subclass N of $M = M(R^n)$, the conclusion of the theorem can be strengthened to

(1) $$\omega_\tau(f;\, a) \leqslant A\omega_\sigma(f;\, Ba)$$

for certain (other) constants A, B depending only on σ and τ. Using this he was able to refine also the conclusion of Theorem 9.4.4.5, in the case that $\sigma \in N$, to

(2) $$\omega_\tau(f;\, a) \leqslant A\, a^r \int_a^\infty u^{-r}\, \omega_\sigma(f;\, u)\, du/u .$$

Concerning the class N, we shall not give its definition here, but only remark that it contains every measure which is the sum of a non-null purely atomic measure and an absolutely continuous one. One very special, but none the less interesting, case of Boman's results is that where $\sigma = \delta - k\, dt$, where δ denotes as usual the Dirac measure (unit point mass at 0), and $k \in L^1(R^n)$, $\int k\, dt = 1$. Here $\hat{\sigma}(x) = 1 - \hat{k}(x) \to 1$ as $|x| \to \infty$, so $\hat{\sigma}$ satisfies the Tauberian condition. Let us show that with this special choice of σ, if $\tau \in M$ and $\hat{\sigma}$ divides $\hat{\tau}$ at 0, then (1) holds.

Indeed, suppose first $\mu \in M$ and $\hat{\mu}(x)$ vanishes for $|x| \leqslant c$. By 7.3.11.2 there exist a number $b > 0$, and $\nu \in M$ such that $\hat{\nu}(x)\, \hat{\sigma}(x) = 1$ for $|x| \geqslant b$. Hence

$$\hat{\mu}(x)(\hat{\nu}(bx/c)\, \hat{\sigma}(bx/c) - 1) = 0 , \quad x \in \hat{R}^n ,$$

i.e. $\hat{\mu}(x) = \hat{\mu}(x)\, \hat{\nu}(bx/c)\, \hat{\sigma}(bx/c)$, showing that μ is a multiple (in M) of $\sigma_{(b/c)}$; hence

(3) $$\omega_\mu(f;\, a) \leqslant A_1\, \omega_\sigma(f;\, B_1 a)$$

holds, with $A_1 = \|\mu\|_M\, \|\nu\|_M$, $B_1 = b/c$. If now $\hat{\sigma}$ divides $\hat{\tau}$ at 0, then, for some $\rho \in M$, $\hat{\mu} = \hat{\tau} - \hat{\rho}\, \hat{\sigma}$ vanishes in a neighborhood of 0. Since

$$\omega_\tau(f;\, a) \leqslant \|\rho\|_M\, \omega_\sigma(f;\, a) + \omega_\mu(f;\, a) ,$$

we obtain (1) upon substituting into this last inequality the estimate (3) for ω_μ. This proves the asserted proposition. Under the same hypothesis on σ, we can similarly deduce the sharpened version of Theorem 9.4.4.5 whereby (2) replaces 9.4.4.5 (1).

Observe that if $\sigma = \delta - k\, dt$, where $\int k\, dt = 1$ and $k(t) \geqslant 0$, then, since $\hat{k}(x) \neq 1$ except at $x = 0$, the hypothesis "$\hat{\sigma}$ divides $\hat{\tau}$ at 0" implies "$\hat{\sigma}$ divides $\hat{\tau}$ (globally!)" in view of 7.3.11.3, and the conclusion of Theorem 9.4.4.4 can be sharpened to

$$\|f * \tau_{(a)}\| \leqslant A\, \|f * \sigma_{(a)}\| ;$$

in particular (1) holds with $B = 1$, in view of 9.4.1.1 (2).

The full results of Boman, which reach much deeper than the special case just discussed, are not yet published. He kindly made available to us a preliminary manuscript and consented to the inclusion of some of his material (which here was placed in the setting of HBS) in the present paragraph.

It should also be mentioned that refinements, of a different character, of Theorems 9.4.4.4 and 9.4.4.5 are known when the underlying HBS is $L^p(R^n)$ or $L^p(T^n)$, $1 < p < \infty$; see BOMAN & SHAPIRO$_{1,2}$.

9.4.5 An "inverse theorem". The preceding comparison theorems have numerous applications; in the case where the underlying HBS is $C(T)$ or $\dot{C}(R)$ (bounded uniformly continuous functions on R), a number of such applications were given in SHAPIRO$_1$; one could repeat these applications nearly word for word in the general context of HBS. Since this is just a matter of routine, we shall choose only one example, to which we have already referred earlier: the inverse theorem (due to S. Bernstein and de la Vallée Poussin) for trigonometric approximation. The classical case is that where $B = C(T)$.

9.4.5.1 Theorem. Let B be a HBS consisting of 2π-periodic measurable functions on R, and containing $\mathcal{T}_n$, $n = 0, 1, \ldots$. Suppose $f \in B$ and $\text{dist}\,(f, \mathcal{T}_n) = O(n^{-c})$, where $0 < c < 1$. Then $\omega(f, a) = O(a^c)$, where ω denotes the B-modulus of continuity of f.

Proof. If τ is the "dipole" measure, which places masses of 1, -1 at the points $t = 0, 1$ respectively, $\omega_\tau(f; a)$ is just $\omega(f, a)$. For this measure $\hat{\tau}(x) = 1 - e^{-ix}$, and τ satisfies the hypotheses of Theorem 9.4.4.5 with $r = 1$, $P(x) = x$. Let now $k \in L^1(R)$, $\hat{k}(x) = 1$ for $|x| \leq 1$, and define σ to be the measure $\delta - k\,dt$. Then, σ satisfies the Tauberian condition. Let now $a > 0$, and denote by $n = n(a)$ the largest integer not exceeding $1/a$. Since $\hat{\sigma}(x)$ vanishes for $|x| \leq 1$, we have, denoting by T_n the closest element to f in $\mathcal{T}_n$, $T_n * \sigma_{(a)} = 0$; hence

$$\|f * \sigma_{(a)}\| = \|(f - T_n) * \sigma_{(a)}\| \leq \|\sigma\|_M \,\text{dist}\,(f, \mathcal{T}_n) = O(a^c) \quad \text{as } a \to 0 .$$

Hence $\omega_\sigma(f; a) = O(a^c)$; substituting this into 9.4.4.5 (1) (with $r = 1$) gives

$$\omega_\tau(f; a) \leqslant A \left[\int_0^a A_1 u^{c-1} du + \int_a^\infty (a/u) A_1 u^{c-1} du \right] \leqslant A_2 a^c$$

and the theorem is proved. ◇

Remark 1. In case $c = 1$, 9.4.4.5 (1) yields

$$\omega_\tau(f; a) = O(a \log (1/a)) .$$

If we use, as our choice of τ, that measure with $\hat{\tau}(x) = (1 - e^{-ix})^2$, then the hypotheses of 9.4.4.5 are satisfied with $r = 2$, $P(x) = x^2$; we now get, in case $c = 1$, the estimate $\omega_\tau(f; a) = O(a)$ from 9.4.4.5 (1). For this τ, however, ω_τ is not a modulus of continuity but rather an analog thereof based upon the second difference (so-called "modulus of smoothness"), and we obtain (in the case $B = C(T)$) a theorem of Zygmund, cf. e.g. LORENTZ$_1$ or NATANSON.

Remark 2. If our hypothesis about f is taken in the more general form

$$\text{dist} (f, \mathcal{T}_n) = O(\psi(n)) ,$$

where $\psi(n)$ is a decreasing function and $\lim \psi(n) = 0$, we can still, as a rule, get information about $\omega(f, a)$ with the aid of Theorem 9.4.4.5. However, when $\psi(n)$ tends very slowly to 0, roughly when $\psi(n)$ tends to zero like $1/\log n$ or slower, the integral on the right of 9.4.4.5 (1), which arises upon application of the above technique, is found to diverge; Theorem 9.4.4.5 fails in this case to give any information. This is a weakness of the theorem, insofar as other methods (cf. LORENTZ$_1$, Section 4.4, or TIMAN, Section 6.1) allow a non-trivial conclusion about $\omega(f, a)$ to be drawn, no matter how slowly $\psi(n)$ tends to zero.

Just this is the point of Boman's refinement of Theorem 9.4.4.5, discussed in 9.4.4.6; the measure σ is of the special type for which 9.4.4.6 (2) holds, and the integral on the right side of that inequality converges always. It can be shown to yield an essentially best possible estimate for $\omega(f, a)$ (corresponding to theorems of A.F. and M.F. Timan when $B = L^p(T)$, $1 \leqslant p < \infty$, and of Stechkin when $B = C(T)$; references in TIMAN); there are a few remarks concerning this matter in BOMAN & SHAPIRO$_2$, and it shall be discussed in detail in a forthcoming work of Boman. It can be shown that for measures σ not of class N (cf. 9.4.4.6), refinements of the Boman type are not possible, i.e. the hypotheses of Theorems 9.4.4.4 and 9.4.4.5 do not allow conclusions as strong as 9.4.4.6 (1) and 9.4.4.6 (2), respectively.

<u>Exercises</u>. a) Let β_1 denote the measure on R^1 consisting of two point-masses, namely $\beta_1(\{0\}) = 1$, $\beta_1(\{1\}) = -1$, and define $\beta_n = \beta_1^n$ (i.e. n-fold convolution power of β_1). For $B = C(T)$, the associated moduli

$$\omega_i(f, a) = \omega_{\beta_i}(f, a)$$

are the <u>moduli of smoothness</u> (of orders $i = 1, 2, \ldots$) of f (ω_1 is, of course, the usual modulus of continuity). Prove, for $i \geq 1$,

$$\omega_i(f, t) \leq C_i\, t^i(\|f\|_\infty + \int_t^A s^{-i-1}\, \omega_{i+1}(f, s)\, ds)$$

where A, C_i are constants independent of f.

b) Prove, for $f \in C(T)$, and $i = 1, 2, \ldots,$

$$\omega_i(f, t) \leq A_i\, t^i \sum_{0 \leq n \leq t^{-1}} (n + 1)^{i-1}\, \text{dist}\,(f, \mathcal{T}_n)$$

where A_i depends only on i.

c) Let $k \in L^1(R)$, $\int k\, dt = 1$, and suppose there exists $\mu \in M(R)$ such that

$$(*) \qquad (1 - \hat{k}(x))\, \hat{\mu}(x) = x^2$$

for x in a neighborhood of 0. Deduce that if $f \in B$ and $\|f - (f * k_{(a)})\| = o(a^2)$ as $a \to 0$, then $\omega_{\beta_2}(f, a) = o(a^2)$. Deduce from this that (i) in case $B = \dot{C}(R)$, f is constant, (ii) in case $B = L^p(R)$, $1 \leq p < \infty$, f is zero. Moreover, characterize those f for which $\|f - (f * k_{(a)})\| = O(a^2)$.

d) Solve the analogous problems when, in place of x^2 on the right hand side of (*), we have x; similarly for $|x|$. (These are special cases of "saturation problems", cf. Appendix I.)

e) Take $B = C(T)$, and let φ denote the function defined in 9.4.2.1. Verify, for any trigonometric polynomial f, that

$$f * \varphi_{(a)} = \tilde{f} * \theta_{(a)}$$

where $\tilde{f}$ is the function conjugate to f (i.e. if $f(t) = \sum_{-n}^{n} a_j e^{ijt}$, then $\tilde{f}(t) = - i \sum_{-n}^{n} (\operatorname{sgn} j)\, a_j e^{ijt}$) and $\hat{\theta}(x) = (i \operatorname{sgn} x)\, \hat{\varphi}(x)$. Use this to deduce <u>Privalov's Theorem</u>, i.e. if $g \in \operatorname{Lip} \alpha$, $0 < \alpha < 1$, then there exists a function in $\operatorname{Lip} \alpha$ whose Fourier coefficient of order n is $(i \operatorname{sgn} n)\, \hat{g}(n)$.

f) Let u be a bounded harmonic function in the upper half plane, satisfying for $y \leq y_0$,

(**) $$\sup_x \left| \frac{\partial u}{\partial x}(x, y) \right| \leq A(y) .$$

(i) If $\int_0^{y_0} A(y)\, dy < \infty$, show u is continuously extendible to the closed half-plane, and obtain an estimate for the modulus of continuity of $u(x, 0)$.

(ii) Construct an example where (**) holds with $A(y) = (y \log 1/y)^{-1}$, yet u is not continuously extendible to the closed half-plane. (Note: this is an example of a problem whereby Theorems 9.4.4.4 and 9.4.4.5 are applicable, cf. SHAPIRO_1, Chapter V, but the refined estimates discussed in 9.4.4.6 are not.)

(iii) Carry out a similar analysis for each of the differential operators $\frac{\partial u}{\partial y}$, $\frac{\partial^2 u}{\partial x^2}$, $\frac{\partial^2 u}{\partial x\, \partial y}$, as well as generalizations to bounded harmonic functions in a half-space of R^n.

BIBLIOGRAPHY

1. AHIEZER N.I., Vorlesungen über Approximationstheorie. Akademie-Verlag, Berlin 1953

2. ——— , The Classical Moment Problem and some related Questions in Analysis. Oliver & Boyd, Edinburgh and London 1965.

ARONSZAJN N., Theory of reproducing kernels. Trans. Amer. Math. Soc. 68 (1950) 337-404.

BANACH S., Théorie des Opérations Linéares. Hafner Publishing Company, New York.

BARGMANN V., On a Hilbert space of analytic functions and an associated integral transform, Part I. Comm. Pure Appl. Math. 14 (1961) 187-214.

BARRAR R.B. & H.L. LOEB, On the continuity of the nonlinear Tschebyscheff operator. Pacific J. Math. 32 (1970) 593-601.

BEREZANSKI Ju. M., Expansions in Eigenfunctions of Selfadjoint Operators. Transl. of Math. Monographs vol. 17. Amer. Math. Soc. Providence R.I. 1968.

BERGMAN S., The Kernel Function and Conformal Mapping. Math. Surveys No. 5, Amer. Math. Soc. New York, N.Y. 1950 (rev. ed. 1970).

BERMAN D.L., On the impossibility of constructing a linear polynomial operator giving an approximation of best order. Dokl. Akad. Nauk SSSR 120 (1958) 1175-1177.

BEURLING A., Sur les intégrales de Fourier absolument convergentes et leur application à une transformation fonctionelle. Neuvième congres des mathematiciens scandinaves, Helsingfors 1938.

BIEBERBACH L., Analytische Fortsetzung. Springer-Verlag, Berlin-Göttingen-Heidelberg 1955.

BOAS R.P., Entire Functions. Academic Press, New York 1954.

1. BOCHNER S., Harmonic Analysis and the Theory of Probability. Univ. of California Press, Berkeley and Los Angeles 1955.

2. ——— , Localization of best approximation. In "Contributions to Fourier Analysis" Annals of Math. Studies 25, Princeton Univ. Press, Princeton 1950, pp. 3-23.

BOCHNER S. & K. CHANDRASEKHARAN, Fourier Transforms. Annals of Math. Stud. no 19, Princeton 1949.

BOHR H. & E. FØLNER, On some types of functional spaces. Acta Math. 76 (1945) 31-155.

BOJANIC R. & R. DeVORE, A proof of Jackson's theorem. Bull. Amer. Math. Soc. 75 (1969) 364-367.

BOMAN J., Partial regularity of mappings between Euclidean spaces. Acta Math. 119 (1967) 1-25.

1. BOMAN J. & H.S. SHAPIRO, Comparison theorems for a generalized modulus of continuity. Bull. Amer. Math. Soc. 75 (1969) 1266-1268.

2. ——— , Comparison theorems for generalized modulus of continuity. To appear in Ark. Mat.

BONSALL F.F., Dual extremum problems in the theory of functions. J. London Math. Soc. 31 (1956) 105-110.

de BRANGES L., Hilbert Spaces of Entire Functions. Prentice-Hall Inc. Englewood Cliffs, N.J. 1968.

1. BROSOWSKI B., Über Extremalsignaturen linearen Polynome in n Veränderlichen. Numer. Math. 7 (1965) 396-405.

2. ——— , Nicht-lineare Tschebyscheff-Approximation. Bibliographisches Institut AG, Mannheim 1968.

1. BUCK R.C., Linear spaces and approximation theory. In "On Numerical Approximation", pp. 11-24. Editor: R.E. Langer. Univ. of Wisconsin Press, Madison 1959.

2. ——— , Applications of duality in approximation. In "Approximation of Functions", pp. 27-42. Editor: H.L. Garabedian. Elsevier Publ. Co. 1965.

3. ——— , Survey of recent Russian literature on approximation. In "On Numerical Approximation", pp. 341-359. Editor: R.E. Langer. Univ. of Wisconsin Press, Madison 1959.

BUTZER P.L. & R.J. NESSEL, Fourier Analysis and Approximation I, II. Birkhäuser Verlag, Basel und Stuttgart 1970-71.

CHALMERS B.L., Subspace kernels and minimum problems in Hilbert spaces with kernel function. Pacific J. Math. 31 (1970) 620-627.

CHENEY E.W., Introduction to Approximation Theory. McGraw-Hill, New York 1966.

CHENEY E.W. & C.R. HOBBY & P.D. MORRIS & F. SCHURER & D.E. WULBERT, On the minimal property of the Fourier projection. Trans. Amer. Math. Soc. 143 (1969) 249-258.

COLLATZ L., Approximation von Funktionen bei einer und bei mehreren unabhängigen Veränderlichen. Z. Angew. Math. Mech. 36 (1956) 198-211.

COURANT R., Dirichlet Principle, Conformal Mapping and Minimal Surfaces. Interscience Publishers, New York 1950.

COURANT R. & D. HILBERT, Methods of Mathematical Physics, I. Interscience Publishers, New York - London 1953.

CURTIS P.C., n-parameter families and best approximation. Pacific J. Math. 9 (1959) 1013-1027.

DAVIS P.J., Interpolation and Approximation. Blaisdell Publishing Co. 1963.

1. DAY M.M., Normed linear spaces. Ergebn. der Math. u. ihrer Grenzgeb. N.F. Heft 21. Springer-Verlag, Berlin-Göttingen-Heidelberg 1962.

2. ——, Some characterizations of inner-product spaces. Trans. Amer. Math. Soc. 62 (1947) 320-337.

DEUTSCH F.R. & P.H. MASERICK, Applications of the Hahn-Banach theorem in approximation theory. SIAM Rev. 9 (1967) 516-530.

DIAZ J.B., Upper and lower bounds for eigenvalues. Proc. of Symp. in Appl. Math. VIII, pp. 53-78. Amer. Math. Soc. Providence, R.I. 1958.

1. DOMAR Y., On the uniqueness of minimal extrapolations. Ark. Mat. 4 (1960) 19-29.

2. DOMAR Y., An extremal problem related to Kolmogoroff's inequality for bounded functions. Ark. Mat. 7 (1969) 433-441.

DONOGHUE Jr. W.F., Distributions and Fourier transforms. Academic Press, New York - London 1969.

DOOB J.L., A minimum problem in the theory of analytic functions. Duke Math. J. 8 (1941) 413-424.

DUFFIN R.J. & J.J. EACHUS, Some notes on an expansion theorem of Paley and Wiener. Bull. Amer. Math. Soc. 48 (1942) 850-855.

DUREN P.L., Theory of H^p Spaces. Academic Press, New York and London 1970.

DUREN P.L. & D. Williams, Some interpolation problems for analytic functions. To appear.

DYM H. & H.P. McKEAN Jr., Application of de Branges spaces of integral functions to the prediction of stationary Gaussian processes. Illinois J. Math. 14 (1970) 299-343.

EDWARDS R.E., Fourier Series: a modern introduction, 2 Vols. Holt, Rinehart and Winston, New York 1967.

ESSEEN C.G., Fourier analysis of distribution functions. Acta Math. 77 (1945) 1-125.

1. FREUD G., Orthogonale Polynome. Birkhäuser Verlag, Basel und Stuttgart 1969.

2. ——, Über die Approximation reeller Funktionen durch rationale gebrochene Funktionen. Acta Math. Acad. Sci. Hungar 17 (1966) 313-324.

FREUD G. & S. KNAPOWSKI, On linear processes of approximation I, II, III. Studia Math. 23 (1963) 105-112, ibidem 25 (1964-1965) 251-263, 373-383.

FREUD G. & J. SZABADOS, Rational approximation to x^α. Acta Math. Acad. Sci. Hungar 18 (1967) 393-399.

GANELIUS T., Lectures on Tauberian Remainder Theorems. Matscience, Madras 1969.

GANELIUS T. & S. Westlund, The degree of approximation in Müntz's theorem. To appear in "Proceedings of the International Conference on Mathematical Analysis", Jyväskylä Finland 1970.

GOLDBERG R.R., Fourier Transforms. Cambridge Univ. Press, New York 1961.

1. GOLOMB M., Lectures on Theory of Approximation. Argonne National Laboratory, Appl. Math. Div. 1962.

2. ——, Optimal and nearly-optimal linear approximation. In "Approximation of Functions", pp. 83-100. Editor: H.L. Garabedian. Elsevier Publ. Co. 1965.

3. ——, Optimal approximating manifolds in L_2-spaces. J. Math. Anal. Appl. 12 (1965) 505-512.

GOLOMB M. & H.F. WEINBERGER, Optimal approximation and error bounds. In "On Numerical Approximation", pp. 117-190. Editor: R.E. Langer. Univ. of Wisconsin Press, Madison 1959.

1. GONCHAR A.A., Estimates of the growth of rational functions and some of their applications. Mat. Sb. (N.S.) 72 (114) (1967) 489-503.

2. ——, Rapidity of rational approximation of continuous functions with characteristic singularities. Mat. Sb. (N.S.) 73 (115) (1967) 630-638.

GOULD S.H., Variational Methods for Eigenvalue Problems. University of Toronto Press, Toronto 1957.

GRENANDER U. & G. SZEGÖ, Toeplitz Forms and their Applications. Univ. of California Press, Berkeley and Los Angeles 1958.

GROSOF M.S. & D.J. NEWMAN, Haar polynomials on Cartesian product spaces. Duke Math. J. 36 (1969) 193-206.

HAAR A., Die Minkowskische Geometrie und die Annäherung an stetige Funktionen. Math. Ann. 78 (1918) 294-311.

HAMBURGER H.L. & M.E. GRIMSHAW, Linear Transformations in n-dimensional Vector Space. Cambridge University Press 1951.

HAMMERSLEY J.M., A non-harmonic Fourier series. Acta Math. 89 (1953) 243-260.

HANNER O., On the uniform convexity of L^p and ℓ^p. Ark. Mat. 3 (1956) 239-244.

1. HAVINSON S. Ja., The theory of extremal problems for bounded analytic functions satisfying additional conditions inside the domain. Uspechi Mat. Nauk 18, (2) (1963) 25-98. Russ. Math. Surv. 18, (2) (1963) 23-96.

2. ——— , On approximation, with reference to the size of the coefficients of the approximants. Trudy Mat. Inst. Steklov 60 (1961) 304-324. Amer. Math. Soc. Transl. (2) 44 (1965) 67-88.

HEINS M., The minimum modulus of a bounded analytic function. Duke Math. J. 14 (1947) 179-215.

HELSON H., Lectures on Invariant Subspaces. Academic Press, New York - London 1964.

HERMES S.H. & J.P. LaSALLE, Functional Analysis and Time Optimal Control. Academic Press, New York and London 1969.

HERTZ C.S., A note on summability methods and spectral analysis. Trans. Amer. Math. Soc. 86 (1957) 506-510.

HEWITT E. & K.A. ROSS, Abstract Harmonic Analysis I, II. Springer-Verlag, Berlin-Göttingen-Heidelberg 1963, 1970.

HILLE E. & R.S. PHILLIPS, Functional Analysis and Semi-Groups. Amer. Math. Soc. Colloquium Publ. 31, Providence R.I. 1957.

HINTZMAN W., Best uniform approximation via annihilating measures. Bull. Amer. Math. Soc. 76 (1970) 1062-1066.

HOFFMAN K., Banach Spaces of Analytic Functions. Prentice Hall Inc., N.J. 1962.

1. HÖRMANDER L., Föreläsningar över distributionsteori (in Swedish). Mat. Inst. Lunds Univ. Sweden.

2. ——— , Estimates for translation invariant operators in L^p spaces. Acta Math. 104 (1960) 93-140.

INCE E.L., Ordinary Differential Equations. Dover Publications, New York 1944.

IOFFE A.D. & V.M. TIHOMIROV, Duality of convex functions and extremum problems. Uspechi Mat. Nauk 23 (6) (1968) 51-116. Russ. Math. Surv. 23 (6) (1968) 53-124.

JACKSON D., The Theory of Approximation. Amer. Math. Soc. Colloquium Publ. XI, New York 1930.

JAMES R.C., Orthogonality and linear functionals in normed linear spaces. Trans. Amer. Math. Soc. 61 (1947) 265-292.

1. JEROME J.W., On n-widths in Sobolev spaces and applications to elliptic boundary value problems. J. Math. Anal. Appl. 29 (1970) 201-215.

2. ——, On the L^2 n-width of certain classes of functions of several variables. J. Math. Anal. Appl. 20 (1967) 110-123.

JEROME J.W. & L.L. SCHUMAKER, Applications of ϵ-entropy to the computation of n-widths. Proc. Amer. Math. Soc. 22 (1969) 719-722.

KAHANE J.P., Series de Fourier absolument convergentes. Springer-Verlag, Berlin-Heidelberg-New York 1970.

KAHANE J.P. & R. SALEM, Ensembles Parfaits et Séries Trigonométriques. Hermann, Paris 1963.

KALLIONIEMI H.H., On inequalities between the uniform bounds on the derivatives of a complex-valued function of a real variable. Inaugural dissertation, Uppsala Univ. (Sweden) 1970.

KARLIN S.J. & W.J. STUDDEN, Tschebycheff systems: with applications in analysis and statistics. Pure and Applied Math. Vol XV, Wiley & Sons, 1966.

KATZNELSON Y., An Introduction to Harmonic Analysis. Wiley & Sons Inc. 1968.

KOLMOGOROV A.N. & V.M. TIHOMIROV, ϵ-entropy and ϵ-capacity of sets in function spaces. Uspechi Mat. Nauk 14 (2) (1959) 3-86. Amer. Math. Soc. Transl. (2) 17 (1961) 277-364.

KOROVKIN P., Linear Operators and Approximation Theory. Fizmatgiz, Moscow 1959 (Russian). Hindustan Publ. Corp., Dehli 1960.

KÖTHE G., Topological Vector Spaces, I. Springer-Verlag, Berlin-Heidelberg-New York 1969.

KREIN M.G., The ideas of P.L. Cebysev and A.A. Markov in the theory of limiting values of integrals and their further development. Uspechi Mat. Nauk 6 no 4 (44) (1951) 3-120. Amer. Math. Soc. Transl. (2) 12 (1959) 1-122.

KRIPKE B.R. & T.J. RIVLIN, Approximation in the metric $L^1(X, \mu)$. Trans. Amer. Math. Soc. 119 (1965) 101-122.

KRYLOV V.I., Approximate Calculation of Integrals. The Macmillan Co., New York 1962.

KUPTSOV N.P., Direct and converse theorems of approximation theory and semigroups of operators. Uspechi Mat. Nauk 23 (4) (1968) 117-179. Russ. Math. Surv. 23 (4) (1968) 115-179.

LAMBERT P., On the minimum norm property of the Fourier projection in L^1-spaces and in spaces of continuous functions. Bull. Amer. Math. Soc. 76 (1970) 798-804.

LANDAU E., Darstellung und Begründung einiger neuerer Ergebnisse der Funktionentheorie. Chelsea Publishing Company, New York 1946.

LANDAU H.J. & H.O. POLLAK, Prolate spheroidal wave functions, Fourier analysis and uncertainty II, III. Bell Syst. Tech. J. 40 (1961) 65-84, ibid. 41 (1962) 1295-1336.

LAWSON C.,Dissertiation at the University of California, Los Angeles 1961.

LAX P.D., Reciprocal extremal problems in function theory. Comm. Pure Appl. Math. 8 (1955) 437-455.

DE LEEUW K. & H. MIRKIL, A priori estimates for differential operators in L^∞ norm. Illinois J. Math. 8 (1964) 112-124.

LEVINSON N., Gap and Density Theorems. Amer. Math. Soc. Colloquium Publ. XXVI, New York 1940.

LEVINSON N. & H.P. McKEAN Jr., Weighted trigonometrical approximation on R^1 with application to the germ field of a stationary Gaussian noise. Acta Math. 112 (1964) 99-143.

LOGAN B. Jr., Properties of High-Pass Signals. Dissertation presented to the Electrical Engineering faculty of Columbia University, 1965.

1. LORENTZ G.G., Approximation of Functions. Holt, Rinehart and Winston, 1966.

2. ——— , Russian literature on approximation in 1958-1964. In "Approximation of Functions", pp. 191-215. Editor: H.L. Garabedian. Elsevier Publ. Co. 1965.

3. ——— , Bernstein Polynomials. Univ. of Toronto Press, Toronto 1953.

LUENBERGER D.G., Optimization by Vector Space Methods. Wiley & Sons, New York 1969

MACINTYRE A.J. & W.W. ROGOSINSKI, Extremum problems in the theory of analytic functions. Acta Math. 82 (1950) 275-325.

MARKOV V., Über die Funktionen, die in einem gegebenen Intervall möglichst wenig von Null abweichen. Math. Ann. 77 (1916) 213-258.

MASANI P., Wiener's contribution to generalized harmonic analysis, prediction theory and filter theory. Bull. Amer. Math. Soc. 72, no. 1, part II (1966) 73-125.

MEINARDUS G., Approximation von Funktionen und Ihre Numerische Behandlung. Springer-Verlag, Berlin 1964 (Engl. ed. 1967).

MESCHKOWSKI H., Hilbertsche Räume mit Kernfunktion. Springer-Verlag, Berlin-Göttingen-Heidelberg 1962.

MIKHLIN S.G., The Problem of the Minimum of a Quadratic Functional. Holden-Day, San Francisco-London-Amsterdam 1965.

MILMAN D.P., On some criteria for the regularity of spaces of the type (B). Dokl. Akad. Nauk SSSR, N.S. 20 (1938) 243-246.

MITYAGIN B.S., Approximation of functions in L^p and C spaces on the torus. Mat. Sbornik (N.S.) 58 (100) (1962) 397-414.

MÜLLER C., Spherical Harmonics. Springer-Verlag, Berlin-Heidelberg-New York 1966.

NACHBIN L., Elements of Approximation Theory. Van Nostrand Math. Studies 14, Princeton N.J. 1967.

NATANSON I.P., Konstruktive Funktionentheorie. Akademie-Verlag, Berlin 1955.

NEWMAN D.J., Efficiency of polynomials on sequences. J. Approximation Theory 1 (1968) 66-76.

1. NEWMAN D.J. & L. RAYMON, A class of curves on which polynomials approximate efficiently. Proc. Amer. Math. Soc. 19 (1968) 595-599.

2. ——, Quantitative polynomial approximation on certain planar sets. Trans. Amer. Math. Soc. 136 (1969) 247-259.

1. NEWMAN D.J. & H.S. SHAPIRO, Some theorems on Cebysev approximation. Duke Math. J. 30 (1963) 673-682.

2. ——, Jackson's theorem in higher dimensions. In "On Approximation Theory", ISNM vol 5, pp. 208-219. Birkhäuser, Basel und Stuttgart 1964.

3. ——, Taylor coefficients of inner functions. Mich. Math. J. 9 (1962) 249-255.

4. ——, Certain Hilbert spaces of entire functions. Bull. Amer. Math. Soc. 72 (1966) 971-977.

5. ——, Fischer spaces of entire functions. Proceedings of Symposia in Pure Math. vol. XI, pp. 360-369. Amer. Math. Soc. Providence, R.I. 1968.

NIKOLSKI S.M., Approximation of Functions of Several Variables and Embedding Theorems. NAUKA, Moscow 1969.

NORDLANDER G., The modulus of convexity in normed linear spaces. Ark. Mat. 4 (1960) 15-17.

OBERHETTINGER F., Tabellen zur Fourier Transformation. Springer-Verlag, Berlin-Göttingen-Heidelberg 1957.

POLYA G. & G. SZEGÖ, Aufgaben und Lehrsätze aus der Analysis II, 2te Auflage. Springer-Verlag, Berlin-Heidelberg-New York 1954.

RAGOZIN D.L., Polynomial approximation on compact manifolds and homogeneous spaces. Trans. Amer. Math. Soc. 150 (1970) 41-53.

RAHMAN Q.I., Applications of Functional Analysis to Extremal Problems for Polynomials. Les Presses de l'Université de Montreal 1968.

REITER H., Classical Harmonic Analysis and Locally Compact Groups. Oxford Univ. Press, Oxford 1968.

RICE J.R., The Approximation of Functions. Volume I: Linear Theory; Volume II: Nonlinear and Multivariate Theory. Addison-Wesley, Reading, Mass. 1964, 1969.

RIESZ F. & SZ.NAGY B., Lecons d'Analyse Fonctionelle. Gauthier-Villars, Paris 1955

RIVLIN T.J., An Introduction to the Approximation of Functions. Blaisdell, Waltham, Mass. 1969.

RIVLIN T.J. & H.S. SHAPIRO, A unified approach to certain problems of approximation and minimization. J. Soc. Indust. and Appl. Math. 9 (1961) 670-699.

RIVLIN T.J. & B. Weiss, Some best polynomial approximations in the plane. Duke Math. J. 35 (1968) 475-482.

ROCKAFELLAR R.T., Convex Analysis. Princeton University Press, Princeton 1970.

ROGOSINSKI W.W. & H.S. SHAPIRO, On certain extremum problems for analytic functions. Acta Math. 90 (1953) 287-318.

ROSENBAUM J.T., Simultaneous interpolation in H^2. Mich. Math. J. 14 (1963) 65-70.

1. RUDIN W., Fourier Analysis on Groups. Interscience Publishers, New York-London 1962

2. ——, Real and Complex Analysis. McGraw-Hill, New York-Toronto-London 1966.

SARD A., Linear Approximation. Math. Surv. No. 9, Amer. Math. Soc. Providence 1963

SCHOENBERG I.J. & C.T. YANG, On the unicity of solutions of problems of best approximation. Ann. Mat. Pure Appl. (4) 54 (1961) 1-12.

1. SHAPIRO H.S., Smoothing and Approximation of Functions. Van Nostrand Reinhold Co., New York 1969.

2. ——, On a class of extremal problems for polynomials in the unit circle. Portugal. Math. 20 (1961) 67-93.

3. ——, Approximation by trigonometric polynomials to periodic functions of several variables. In "Abstract Spaces and Approximation", ISNM 10, pp. 203-217. Birkhäuser, Basel und Stuttgart 1969.

4. ——, Some theorems on Cebysev approximation II. J. Math. Anal. Appl. 17 (1967) 262-268.

5. SHAPIRO H.S., Reproducing kernels and Beurling's theorem. Trans. Amer. Math. Soc. 110 (1964) 448-458.

6. ——, Some function-theoretic problems motivated by the study of Banach algebras. To appear in "Proceedings of the Conference on Classical Function Theory", Washington D.C. 1970.

7. ——, Boundary values of holomorphic functions of several variables. To appear in Bull. Amer. Math. Soc.

8. ——, Some negative theorems of approximation theory. Michigan Math. J. 11 (1964) 211-217.

9. ——, A power series with small partial sums. Notices Amer. Math. Soc. 5 (1958) 366.

1. SHAPIRO H.S. & A.L. SHIELDS, On some interpolation problems for analytic functions. Amer. J. Math. 83 (1961) 513-532.

2. ——, On the zeros of functions with finite Dirichlet integral and some related function spaces. Math. Z. 80 (1962) 217-229.

1. SINGER I., Best Approximation in Normed Linear Spaces by Elements of Linear Subspaces. Springer-Verlag, Berlin-Heidelberg-New York 1970.

2. ——, Bases in Banach Spaces I. Springer-Verlag, Berlin-Heidelberg-New York 1970.

SLEPIAN D. & H.O. POLLAK, Prolate spheroidal wave functions, Fourier analysis and uncertainty I. Bell Syst. Tech. J. 40 (1961) 43-63.

1. STECHKIN S.B., Best approximation of linear operators. Mat. Zamyetki 1, no 2 (1967) 137-148 (Russian).

2. ——, (editor), Approximation of Functions in the Mean. Proc. of the Steklov Inst. of Math. nr 88 (1967). Amer. Math. Soc. Providence R.I. 1969.

STEIN E.M. & G. WEISS, Introduction to Fourier Analysis on n-dimensional Euclidean Space. Princeton Univ. Press. 1970.

STENGER W., Nonclassical choices in variational principles for eigenvalues. J. Functional Anal. 6 (1970) 157-164.

STOER J. & C. WITZGALL, Convexity and Optimization in Finite Dimensions I. Springer-Verlag, Berlin-Heidelberg-New York 1970.

SZABADOS J., Rational approximation in certain classes of functions. Acta Math. Acad. Sci. Hungar. 19 (1968) 81-85.

SZEGÖ G., Orthogonal Polynomials. Amer. Math. Soc. Colloquium Publ. 23, Providence R.I. 1959.

1. SZ.NAGY B., Über gewisse Extremalfragen bei transformierten trigonometrischen Entwicklungen II. Ber. Math.-Phys. Kl. Sächs. Akad. Wiss. Leipzig 91, 1939.

2. ——, Expansion theorems of Paley-Wiener type. Duke Math. J. 14 (1947) 975-978.

TALBOT A. (editor), Approximation Theory. Proc. of a Symp., Lancaster 1969. Academic Press, London and New York 1970.

1. TIHOMIROV V.M., Widths of sets in functional spaces and the theory of best approximations. Uspechi 15, no 3 (1960) 81-120. Russ. Math. Surv. 15, no 3 (1960) 75-112.

2. ——, Some problems in approximation theory. Dokl. Mat. Nauk 160 (1965) 774-777. Soviet Math. 6 (1965) 202-205.

TIMAN A.F., Theory of Approximation of Functions of a Real Variable. Pergamon Press-Macmillan, New York 1963.

TITCHMARSH E.C., An Introduction to the Theory of Fourier Integrals. Oxford University Press, New York 1948.

1. TRICOMI F.G., Vorlesungen über Orthogonalreihen. Springer-Verlag, Berlin-Göttingen-Heidelberg 1955.

2. ——, Integral Equations. Interscience Publishers, New York - London 1957.

TURAN P. & P. SZÜSZ, On the constructive theory of functions, III. Studia Sci. Math. Hungar. 1 (1966) 315-322.

VITUSHKIN A.G., Estimation of the difficulty of the tabulation problem. Gosizdat, Moscow 1959. Engl. ed: Theory of Transmission and Processing of Information. Pergamon Press, New York 1966.

VORONOVSKAYA E.V., The Method of Functionals and its Applications. Transl. of Math. Monographs 28, Amer. Math. Soc. 1969.

1. WALSH J.L., Interpolation and Approximation by Rational Functions in the Complex Domain. Amer. Math. Soc. Colloquium Publ. XX, 2nd ed., 1956.

2. —— , Approximation by rational functions: open problems. J. Approximation Theory 3 (1970) 236-242.

WARREN H.E., Lower bounds for approximation by nonlinear manifolds. Trans. Amer. Math. Soc. 133 (1968) 167-178.

WERNER H., Vorlesungen über Approximationstheorie. Springer-Verlag, Berlin-Heidelberg-New York 1966.

ZUHOVITSKI S.I., On the approximation of real functions in the sense of P.L. Tchebycheff. Uspechi Mat. Nauk 11, no. 2 (68) (1956) 125-159. Amer. Math. Soc. Transl. (2) 19 (1962) 221-252.

ZYGMUND A., Trigonometric Series, 2 vols. Cambridge University Press 1959.

APPENDIX I.

Saturation Problems and Distribution Theory

by

Jan Boman

1. Introduction. Denote by $x \to W(t, x, f)$ for $t > 0$ some approximation to the function $f \in L_p(R^n)$. A typical example is $W(t, x, f)$ defined by

$$W(t, x, f) = t^{-n} \int k(y/t)\, f(x - y)\, dy$$

for almost every $x \in R^n$ and $t > 0$, where $k \in L_1(R^n)$. The saturation problem of the approximation process $W(t, x, f)$ is concerned with finding a function $\alpha(t)$ defined for $t > 0$ such that (a) and (b) below hold ($\| \|_p$ denotes the L_p-norm):

(a) the condition

$$\|W(t, \cdot, f) - f(\cdot)\|_p = o(\alpha(t)), \text{ as } t \to 0$$

is satisfied only by functions f belonging to some given class T, called the trivial class (in most cases T consists only of the null-function);

(b) the condition

$$\|W(t, \cdot, f) - f(\cdot)\|_p = O(\alpha(t)), \text{ as } t \to 0$$

is satisfied by a sufficiently rich set of functions (in general a dense set of functions in L_p).

The function $\alpha(t)$ is called the saturation order of the approximation process in question, and the class of functions satisfying (b) is called the saturation class. If $\alpha(t) = t^\gamma$ for some $\gamma > 0$, one also says that the saturation order is γ. Of course the function $\alpha(t)$ is far from being uniquely determined by (a) and (b). Moreover, there does not always exist a function $\alpha(t)$ such that (a) and (b) hold.

The problem of saturation has been treated in several articles during the last few years. For instance, P.L. Butzer, R.J. Nessel and H. Berens devoted a series of five papers [3], [4], [5], [6], [7] to the subject. Other recent articles are those of E. Görlich [8], [9], [10] and one article by M. Kozima and G. Sunouchi [12].

The purpose of this note is first of all to demonstrate that the proofs of general saturation theorems can be formulated very simply indeed if one makes use of the proper fundamentals of distribution theory. In this way one obtains at one stroke a quite general result for L_p-norms, $1 \leq p \leq \infty$. The general theorem is found in Section 3, its proof in Section 4, and examples are given in Section 5.

A second object of this note is to formulate sufficient conditions for standard type saturation which are very simple to apply to specific kernels. Such conditions are given in Corollaries 3, 4 and 5. The condition of Corollary 4, especially, is very simple to apply, although it is in fact general enough to be satisfied by most of the kernels that have been studied in the literature. To apply the condition to a specific kernel one need only check that the kernel is radial, that it positive, and that it is sufficiently small at infinity.

A few words on previous treatments of the saturation problem. Butzer and Nessel, who completely avoid distribution theory, prove in [3] approximately the same result as the theorem given below, although only for L_p-norms with $1 \leq p \leq 2$. In a later article, [7], Berens and Nessel extend the result in [3] to the case $2 < p < \infty$ by means of a so-called dual method. Kozima and Sunouchi consider in [12] only saturation of order γ where γ is a positive integer. These authors use distribution theory to some extent. Görlich, who also uses distribution theory, treats in [10] saturation of order γ for arbitrary positive γ. The first to point out the usefulness of distribution theory in the study of saturation problems was H. Buchwalter [1].

The reason for the usefulness of distribution theory in saturation problems is easily seen by considering a simple example. Let us consider the approximation of a function $f \in L_p(R)$ by its translate $f_t : x \to f(x + t)$. This approximation process is saturated of order 1, which means in particular that for $1 \leq p < \infty$

$$(1.1) \qquad \|f - f_t\|_p = o(t), \quad \text{as } t \to 0$$

implies

$$f = 0 .$$

To prove this simple assertion using distribution theory we proceed as follows. It is easily seen that for any $f \in L_p$

$$(f_t - f)/t$$

converges as $t \to 0$ in the distribution sense (e.g. in the $\mathcal{S}'$-topology) to the distributional derivative of f, which we denote by Df. On the other hand, it follows from (1.1) that

$$\|(f_t - f)/t\|_p \to 0, \quad \text{as } t \to 0 .$$

Since L_p-convergence implies $\mathcal{S}'$-convergence and the $\mathcal{S}'$-limit is unique, these facts imply that Df = 0, i.e. that Df is equal to the null-distribution. Finally, using the theorem of distribution theory which states that Df = 0 implies f = constant, we obtain the desired conclusion.

This argument can be said to depend on an asymptotic expansion of $f_t - f$ of the form

$$f_t - f = t \cdot Df + \ldots,$$

where the omitted term is o(t) in a sense which could easily be made precise. Note that Df can be written as a convolution D * f, where D is a distribution, the derivative of the Dirac measure with reversed sign. More generally, the saturation properties of the approximation process W(t, x, f) depend on an asymptotic expansion

$$W(t, \cdot, f) - f = t^{\gamma} \cdot T * f + o(t^{\gamma}), \quad \text{as } t \to 0 , \tag{1.2}$$

where T is some tempered distribution and γ is some positive number. If the expression $o(t^{\gamma})$ is interpreted in a proper way, it follows immediately from (1.2) (compare the argument above) that

$$\|W(t, \cdot, f) - f(\cdot)\|_p = o(t^{\gamma}), \quad \text{as } t \to 0 ,$$

if and only if $\quad T * f = 0 .$

In the same way it follows that

$$T * f \in L_p$$

implies $$\|W(t, \cdot, f) - f(\cdot)\|_p = O(t^\gamma), \text{ as } t \to 0 .$$

In case W(t, x, f) is of the form

$$W(t, x, f) = t^{-n} \int k(y/t)\, f(x - y)\, dy = K_t * f(x) ,$$

where $K_t(x) = t^{-n} k(x/t)$ and $k \in L_1(R^n)$, then all the terms of (1.2) are convolutions with f, so (1.2) can be replaced by

(1.3) $$K_t - \delta = t^\gamma T + o(t^\gamma), \text{ as } t \to 0 .$$

Here δ denotes the Dirac measure. Taking Fourier transforms (in the sense of the theory of distributions) we obtain

(1.4) $$\hat{k}(t\xi) - 1 = t^\gamma \hat{T}(\xi) + o(t^\gamma), \text{ as } t \to 0 .$$

If $\hat{k}(\xi)$ has an expansion in a neighborhood of the origin of the form

(1.5) $$\hat{k}(\xi)-1 = C(|\xi|^\gamma + o(|\xi|^\gamma)), \text{ as } |\xi| \to 0 ,$$

then one obtains (1.4) with $\hat{T}(\xi)= C\,|\xi|^\gamma$. Note that since $k \in L_1$, the distribution $\hat{k}(\xi)$ is actually equal to a continuous function. If $\hat{k}(\xi)$ is sufficiently differentiable at the origin, then (1.5) is just a Taylor expansion. So, in this case, the expansion (1.3) can be viewed as a transformed version of the Taylor expansion of $\hat{k}(t\xi)$ at the origin. One object of the theory is to give a rigorous treatment of the error term. This amounts basically to improving (1.5) to

(1.6) $$\hat{k}(\xi) - 1 = C|\xi|^\gamma(1 + \rho(\xi)) ,$$

where $\rho(\xi)$ is an L_1-transform vanishing at the origin.

We wish also to make here a remark on a technical matter which is connected with the characterization of the saturation class. Several authors (e.g. Görlich in [9] and [10]) prefer using the mollified weight function $(1 + |\xi|^2)^{\gamma/2}$ to the function $|\xi|^\gamma$ in defining the function spaces in question since the latter function has a singularity at the origin. More exactly, this singularity appears to give rise to difficulties in defining the distribution

(1.7) $$|\xi|^{\gamma} \hat{f}(\xi)$$

if $f \in L_p$ for $p > 2$. The difficulty is due to the fact that if $f \in L_p$ and $p > 2$, $\hat{f}(\xi)$ is not necessarily a function (but it is of course always a distribution). However, as will be shown below, the distribution (1.7) can be defined as a tempered distribution, and hence we can work with the weight function $|\xi|^{\gamma}$ in the full range $1 \leqslant p \leqslant \infty$.

2. Notation and preliminaries. We will denote by $\mathcal{S} = \mathcal{S}(R^n)$ the Schwartz class of complex-valued infinitely differentiable rapidly decreasing functions, equipped with the usual topology (see [15]). The class of tempered distributions, which is the dual of $\mathcal{S}$, is denoted $\mathcal{S}'$. The Lebesgue spaces $L_p = L_p(R^n)$ and the space $M = M(R^n)$ of measures with finite total variation are identified with subsets of $\mathcal{S}'$, and so are the Fourier transforms of these spaces, $\hat{L}_p$ and $\hat{M}$ respectively. The Fourier transform of the distribution $f \in \mathcal{S}'$ is denoted $\hat{f}$. For $f \in L_1(R^n)$ the Fourier transform of f can be written

$$\hat{f}(\xi) = \int f(x)\, e^{-i\langle x,\xi\rangle}\, dx, \quad \xi \in R^n ,$$

where $\langle x, \xi\rangle = x_1\xi_1 + \dots + x_n\xi_n$. The norms in L_p and M are denoted $\| \|_p$ and $\| \|_M$, respectively. The corresponding norms on the Fourier transforms of these spaces are denoted $\| \|_{\hat{L}_p}$ and $\| \|_{\hat{M}}$, respectively. We shall frequently use the fact that if $k \in M$ and $f \in L_p$, then $k * f \in L_p$, and

$$\|k * f\|_p \leqslant \|k\|_M \, \|f\|_p .$$

We shall denote by $C^m = C^m(R^n)$ the set of m-times continuously differentiable functions defined in R^n. We denote by $[t]$ the integral part of the real number t. $\overline{C}$ will sometimes denote the set of complex numbers.

The convolution $f * g$ of two distributions $f, g \in \mathcal{S}'$ cannot always be defined as an element of $\mathcal{S}'$ (see [15], ch. VI). If one of f and g has compact support, $f * g$ can be defined. Below we shall form the convolution $Q * f$, where $f \in L_p$ and Q is a distribution in $\mathcal{S}'$ whose Fourier transform $\hat{Q}(\xi) = q(\xi)$ is a positive-homogeneous function of degree $\gamma > 0$ (i.e. $q(t\xi) = t^{\gamma} q(\xi)$ for $t \geqslant 0$ and $\xi \in R^n$) and

$q \in C^{[n/2]+1}(R^n \setminus \{0\})$. We will prove here that the convolution is well-defined in this case. We need a special case of the following lemma.

Lemma 1. (a) Assume that the function $h \in C^N(R^n)$, where $N = [n/2] + 1$, and that there exist constants C and $\delta > 0$ such that an arbitrary derivative $D^m h$ of h of degree m satisfies

$$(2.1) \qquad |D^m h(\xi)| \leq C\, |\xi|^{-\delta - m}, \quad \xi \in R^n, \quad 0 \leq m \leq N .$$

Then $h \in \hat{L}_1(R^n)$.

(b) Assume that $h \in C^N(R^n \setminus \{0\})$, that h has compact support, and that there exist constants C and $\delta > 0$ such that

$$(2.2) \qquad |D^m h(\xi)| \leq C\, |\xi|^{\delta - m}, \quad \xi \in R^n \setminus \{0\}, \; 0 \leq m \leq N .$$

Then $h \in \hat{L}_1(R^n)$.

The conditions of the lemma are satisfied if h has a certain homogeneity property. For instance, if $h \in C^N(R^n)$, and if h outside some compact set is equal to a function which is positive-homogeneous of degree $\delta < 0$, then h satisfies (2.1). This special case of the lemma is given by de Leeuw and Mirkil in [13]. Much sharper statements similar to the lemma can also be found in the literature (see e.g. J. Löfström [14], Corollaries 4.1 and 4.2).

Proof of Lemma 1. We prove proposition (b) here. The proof of (a) is similar. We use a partition of unity of the type

$$\Psi(\xi) = \sum_{k=0}^{\infty} \varphi(2^k \xi) = 1 \quad \text{for} \quad 0 < |\xi| < 1 ,$$

where $\varphi \in C^\infty(R^n)$ and $\varphi(\xi) = 0$ for $|\xi| < 1/2$ and for $|\xi| > 2$. A function φ with these properties can be constructed by taking a function $\lambda \in C^\infty(R)$ such that $\lambda(t) = 1$ for $t < 1$ and $\lambda(t) = 0$ for $t > 2$ and setting $\varphi(\xi) = \lambda(|\xi|) - \lambda(2|\xi|)$. It is sufficient to prove that $\Psi h \in \hat{L}_1$. To estimate the $\hat{L}_1$-norm of

$$h_k : \xi \rightarrow \varphi(2^k \xi)\, h(\xi)$$

one observes that that norm is equal to the $\hat{L}_1$-norm of

$$H_k : \xi \to \varphi(\xi)\, h(2^{-k}\xi) .$$

To estimate $\|H_k\|_{\hat{L}_1}$ one uses the well-known fact that a function $g \in C^N(R^n)$, $N = [n/2] + 1$, whose support is contained in a compact set K, belongs to $\hat{L}_1$, and that there holds an estimate

$$\|g\|_{\hat{L}_1} \leq C \sup \left\{ |D^m g(\xi)| \; ; \xi \in R^n, \; 0 \leq m \leq N \right\} , \tag{2.3}$$

where C depends only on K and n. Computing the derivatives of H_k by means of Leibnitz' formula and using the assumption (2.2) and the inequality (2.3) we obtain

$$\|H_k\|_{\hat{L}_1} \leq C_1 \, 2^{-k\delta} , \; k = 0, 1, 2, \ldots \; .$$

This completes the proof of the lemma.

Assume again that q is a function defined in R^n which is positive-homogeneous of degree $\gamma > 0$ and belongs to the class $C^{[n/2]+1}(R^n \setminus \{0\})$. To define the convolution $Q * f$, where $\hat{Q} = q$ and $f \in L_p(R^n)$, we set

$$\hat{Q}_0(\xi) = w(\xi)\, \hat{Q}(\xi) ,$$

and $Q = Q_0 + Q_1$, where w is an infinitely differentiable function which is equal to 1 in a neighborhood of the origin and equal to 0 in a neighborhood of infinity. Then $Q_0 \in L_1$ by Lemma 1, part (b). Hence $Q_0 * f \in L_p$ for any $f \in L_p$. To consider $Q_1 * f$ we take an integer $m > \gamma/2$ and define a function k by

$$\hat{k}(\xi) = \hat{Q}_1(\xi)/|\xi|^{2m} .$$

Then $k \in \hat{L}_1$ by Lemma 1, part (a), and hence $k * f \in L_p$ for $f \in L_p$. But $Q_1 * f$ is obtained by applying the Laplace operator m times to $k * f$. Since $\mathcal{S}'$ is closed under differentiation, this shows that $Q * f$ is a well-defined element of $\mathcal{S}'$. (The argument actually shows that $Q * f$ belongs to the space $\mathcal{D}'_{L_1}$, which is introduced in [15], ch. VI, § 8.)

Next we introduce some spaces of functions. It is customary to denote by I_γ

the distribution in $\mathcal{S}'(R^n)$ whose Fourier transform is $|\xi|^\gamma$, and by J_γ the distribution whose Fourier transform is $(1 + |\xi|^2)^{\gamma/2}$. Define the space H_p^γ for $1 \leq p \leq \infty$ and $\gamma > 0$ by

$$H_p^\gamma = \left\{f;\ f \in L_p(R^n),\ I_\gamma * f \in L_p(R^n)\right\} .$$

That the convolution $I_\gamma * f$ is well-defined follows from the considerations above. It is easy to prove that the convolution $J_\gamma * f$ is also well-defined for $f \in L_p$ and that for $f \in L_p$ the conditions $I_\gamma * f \in L_p$ and $J_\gamma * f \in L_p$ are equivalent. The last assertion follows from the identities

$$(1 + |\xi|^2)^{\gamma/2} = h_1(\xi) + |\xi|^\gamma h_2(\xi) ,$$

and

$$|\xi|^\gamma = (1 + |\xi|^2)^{\gamma/2} h_3(\xi) ,$$

where $h_j \in \hat{M}$, $j = 1, 2, 3$. The fact that $h_j \in \hat{M}$ can easily be proved by means of Lemma 1. For example, the function h_3 satisfies near the origin the conditions of Lemma 1 part (b) with $\delta = \gamma$, and $1 - h_3$ satisfies at infinity the conditions of Lemma 1 part (a) with $\delta = 2$.

Finally we introduce for $\gamma > 0$ the space

$$M^\gamma = \left\{f;\ f \in L_1(R^n),\ I_\gamma * f \in M(R^n)\right\} .$$

Since for γ an even integer I_γ is an iterate of the Laplace operator, it is obvious that for such γ and $1 \leq p \leq \infty$ the space H_p^γ contains the well-known Sobolev space

$$W_p^\gamma = \left\{f;\ f \in L_p(R^n),\ D^m f \in L_p(R^n) \text{ for } 0 \leq m \leq \gamma\right\} .$$

Here D^m denotes an arbitrary derivative of degree m. Differentiation is of course to be interpreted in the distribution sense. However, if $1 < p < \infty$, the spaces H_p^γ and W_p^γ are identical. This statement is in fact true for arbitrary positive integers γ and $1 < p < \infty$. This is a rather deep result; it follows for instance from the multiplier theorem of Mikhlin (see e.g. Theorem 2.5 in [11]). These results show that the spaces H_p^γ constitute an extension of the Sobolev spaces W_p^γ to non-integral γ.

A more general extension of this kind is furnished by the Besov spaces $B_p^{\gamma,q}$, which depend on three parameters. It is known that

$$H_p^\gamma = B_p^{\gamma,p}$$

for arbitrary real $\gamma > 0$ and $1 < p < \infty$. Many equivalent characterizations of the Besov spaces are known (see e.g. the survey article by Burenkov [2]). Using these results one obtains various equivalent characterizations of the saturation classes in approximation problems.

3. <u>The general saturation theorem.</u> Assume that for each $t > 0$ the function $R^n \rightarrow \overline{C}$

$$x \rightarrow K(t, x)$$

is an element of $L_1(R^n)$ such that

$$\int K(t, x)\, dx = 1 .$$

As an approximation to $f \in L_p(R^n)$ we shall consider the function

$$x \rightarrow W(t, x, f) = \int K(t, x - y)\, f(y)\, dy ,$$

which is defined a.e. in R^n. (We could just as well let $K(t, x)$ be a measure in $M(R^n)$ for each t; this would cause no difficulty except in notation.) An important particular case is

$$K(t, x) = t^{-n}\, k(x/t) ,$$

where $k \in L_1(R^n)$ and $\int k(x)\, dx = 1$.

We denote by $\hat{K}(t, \xi)$ the Fourier transform of $K(t, x)$ with respect to x. As was indicated in the introduction, the saturation phenomena of the approximation process in question are determined basically by the rate by which $\hat{K}(t, \xi) - 1$ converges to 0 as $t \rightarrow 0$.

We shall denote by $\alpha(t)$ a continuous, strictly increasing function defined for $t \geqslant 0$ such that $\alpha(0) = 0$. Let $q(\xi)$ be a continuous function defined in R^n which is positive-homogeneous of degree $\gamma > 0$.

We shall consider the following condition on $K(t, x)$:

$$\hat{K}(t, \xi) - 1 = \alpha(t)\, q(\xi)[1 + \rho(t, \xi)],\ t > 0,\ \xi \in R^n, \tag{3.1}$$

where $\rho(t, \xi) \to 0$ as $t \to 0$ for each fixed ξ. Furthermore we shall assume that $\rho(t, \xi) \in \hat{M}(R^n)$ for each fixed t, and that

$$\|\rho(t, \cdot)\|_{\hat{M}} \leq B < \infty,\ 0 < t < t_0 . \tag{3.2}$$

<u>Theorem.</u> <u>Assume that the kernel</u> $K(t, x)$, $t > 0$, $x \in R^n$, <u>satisfies (3.1) and (3.2), where</u> $q(\xi)$ <u>is positive-homogeneous of degree</u> $\gamma > 0$ <u>and</u> $q \in C^{[n/2]+1}(R^n \setminus \{0\})$. <u>Let</u> Q <u>be the distribution in</u> $\mathcal{S}'$ <u>whose Fourier transform is equal to</u> q. <u>Let</u> $1 \leq p \leq \infty$. <u>Then the following statements hold:</u>

A. $$f \in L_p \text{ and } \|W(t, \cdot, f) - f(\cdot)\|_p = o(\alpha(t)),\ \text{as } t \to 0,$$

<u>implies</u> $$Q * f = 0 .$$

B. (i) $$f \in L_p \text{ and } \|W(t, \cdot, f) - f(\cdot)\|_p = O(\alpha(t)),\ \text{as } t \to 0,$$

<u>if and only if</u>

(ii) $$\begin{cases} f \in L_p \text{ and } Q * f \in L_p \text{ if } 1 < p \leq \infty \\ f \in L_1 \text{ and } Q * f \in M \text{ if } p = 1 . \end{cases}$$

The proof of the theorem is given in the next section.

<u>Remark.</u> The assumption (3.2) in its full strength is used only in the proof of B (in (ii) $\Rightarrow$ (i)). For the proof of A we may replace (3.2) by the weaker condition that

$$|\rho(t, \xi)| \leq C(1 + |\xi|)^K,\ t > 0,\ \xi \in R^n, \tag{3.3}$$

for some real numbers C and K. Note that (3.2) implies (3.3) for $K = 0$.

<u>Corollary 1.</u> <u>Assume that</u> $K(t, x) = t^{-n} k(x/t)$, <u>where</u> k <u>is a function in</u> $L_1(R^n)$ <u>(or a measure in</u> $M(R^n)$<u>) such that</u>

$$\hat{k}(\xi) - 1 = q(\xi)(1 + \rho(\xi)),\ \xi \in R^n,$$

where $q(\xi)$ *is a positive-homogeneous function of degree* $\gamma > 0$ *and*

$$q \in C^{[n/2]+1}(R^n \setminus \{0\}) ,$$

and $\rho(\xi)$ *is a function in* $\hat{M}$ *such that* $\rho(0) = 0$. *Then the assertions A and B of the theorem hold with* $\alpha(t) = t^\gamma$ *and* $Q \in \mathcal{S}'$ *determined by* $\hat{Q}(\xi) = q(\xi)$.

Proof. It is obvious that $\hat{K}(t, \xi) = \hat{k}(t\xi)$ satisfies all assumptions of the theorem with $\alpha(t) = t^\gamma$.

Remark. If $q(\xi) \neq 0$ for $|\xi| \neq 0$, then the assumption in Corollary 1 that $\rho \in \hat{M}$ can be replaced by the apparently weaker assumption that $\rho(\xi)$ belongs locally to $\hat{M}$ in some neighborhood V of the origin. To prove this, let $w(\xi)$ be an infinitely differentiable function equal to 1 in a neighborhood of the origin and equal to 0 outside some compact set contained in the interior of V. Then it follows immediately from the assumptions that

$$w(\xi)\, \rho(\xi) \in \hat{M} . \tag{3.4}$$

Also, since $q(\xi) \neq 0$ for $|\xi| \neq 0$, Lemma 1 shows that

$$q(\xi)^{-1}(1 - w(\xi)) \in \hat{M} ,$$

and hence

$$(1 - w(\xi))\, \rho(\xi) = (1 - w(\xi))(q(\xi)^{-1}(\hat{k}(\xi) - 1) - 1) \in \hat{M} . \tag{3.5}$$

Combining (3.4) and (3.5) we conclude that $\rho \in \hat{M}$.

If $q(\xi) \neq 0$ for $|\xi| \neq 0$ and $p < \infty$, then $f \in L_p$ and $Q * f = 0$ imply $f = 0$. For in this case $\hat{Q}\hat{f} = q\hat{f} = 0$ implies that $\hat{f}(\xi) = 0$ for $|\xi| \neq 0$, i.e. $\hat{f}$ is a distribution with support at the origin. This in turn implies that f is a polynomial. Since $f \in L_p$ and $p < \infty$, it follows that $f = 0$. Similarly, if $q(\xi) \neq 0$ for $|\xi| \neq 0$, then $f \in L_\infty$ and $Q * f = 0$ imply that f is constant.

In most applications $q(\xi)$ is of the form $c|\xi|^\gamma$. It may, however, be useful to consider more general $q(\xi)$ as we have done above. For example, let $n \geq 2$ and let $K(t, x)$ be the measure concentrated on the line $x_2 = \ldots = x_n = 0$ with one-dimensional density

$$(2\pi)^{-1/2}\, t^{-1}\, e^{-x_1^2/2t^2}.$$

Then $$1 - \hat{K}(t, \xi) = 1 - \exp(-t^2\xi_1^2/2) = (1/2)t^2\xi_1^2(1 + \rho(t\xi_1)),$$

where $\rho \in \hat{M}(R)$. Hence the theorem (or Corollary 1) is applicable with $\alpha(t) = t^2$ and $q(\xi) = \xi_1^2/2$.

Choosing $q(\xi) = c|\xi|^\gamma$ in Corollary 1 and taking into account the definition of the spaces H_p^γ and M^γ we obtain

Corollary 2. *Assume that* $K_t(x) = K(t, x) = t^{-n} k(x/t)$, $k \in L_1(R^n)$, *and that*

$$1 - \hat{k}(\xi) = c|\xi|^\gamma (1 + \rho(\xi)),$$

where $c \neq 0$, $\gamma > 0$, $\rho(\xi)$ *belongs locally to* $\hat{M}(R^n)$ *in some neighborhood of the origin, and* $\rho(0) = 0$. *Let* $1 \leqslant p \leqslant \infty$. *Then the following statements hold:*

A. $f \in L_p$ *and* $\|K_t * f - f\|_p = o(t^\gamma)$, *as* $t \to 0$ *implies* $f = 0\,(p < \infty)$ *or* f *constant* $(p = \infty)$, *respectively;*

B. $f \in L_p$ *and* $\|K_t * f - f\|_p = O(t^\gamma)$, *as* $t \to 0$

if and only if $$\begin{cases} f \in H_p^\gamma, \text{ if } 1 < p \leqslant \infty, \\ f \in M^\gamma, \text{ if } p = 1. \end{cases}$$

It must be noted that the assumption

$$1 - \hat{k}(\xi) = c|\xi|^\gamma(1 + \rho(\xi)),$$

where $\lim_{\xi \to 0} \rho(\xi) = 0$, does not suffice to force the conclusion of Corollary 2; it must be assumed also that the remainder term $\rho(\xi)$ is locally a function in $\hat{M}$ (or, equivalently, in $\hat{L}_1$). We will show three examples of additional conditions on $k(x)$ which imply that $\rho(\xi) \in \hat{M}$. The first kind of condition — a differentiability condition on $\hat{k}(\xi)$ — is of course applicable only when γ is an even integer.

Corollary 3. *Assume that* $K_t(x) = t^{-n} k(x/t)$, *where* $k \in L_1(R^n)$ *and for some* $c \neq 0$,

$$1 - \hat{k}(\xi) = c|\xi|^{\gamma}(1 + o(1)), \text{ as } |\xi| \to 0 . \tag{3.6}$$

Assume moreover that γ is an even positive integer and that $\hat{k}$ is a function of class $C^{[n/2]+\gamma+1}$ in some neighborhood of the origin. Let $1 \leq p \leq \infty$. Then the statements A and B of Corollary 2 are valid.

Remark. One can construct examples showing that the differentiability assumptions of Corollary 3 are close to the weakest possible.

Proof. Set

$$\rho(\xi) = c^{-1}\ |\xi|^{-\gamma}\ (1 - \hat{k}(\xi)) - 1 .$$

It follows from the assumptions that ρ is of class $C^{[n/2]+1}$ in some neighborhood of the origin, and this in turn implies that $\rho(\xi)$ is equal to some element of $\hat{L}_1$ in a neighborhood of the origin. Hence the assertion follows from Corollary 2.

We next formulate a condition which is quite simple to apply to specific kernels.

Corollary 4. Assume that $K_t(x) = t^{-n}\ k(x/t)$, where k is a non-negative radial function with integral 1 defined in R^n, whose moments of order $\leq n/2 + 3$ are finite. Then A and B of Corollary 2 hold with $\gamma = 2$.

Remark. If k is assumed to be a measure instead of a function, we must of course add the hypothesis that k is not concentrated at the origin.

Proof. Since $\hat{k}$ clearly belongs to $C^{[n/2]+3}$, we need only show that (3.6) holds with $\gamma = 2$. And this is immediately seen as follows. For reasons of symmetry, all first order moments of k are zero. And since k is also non-negative, all second order moments $\int x_i^2\ k(x)\ dx$ must be non-zero and equal. This shows that $1 - \hat{k}(\xi)$ must have the Taylor expansion

$$c|\xi|^2(1 + o(1)),\ |\xi| \to 0 ,$$

where $c \neq 0$. The proof is complete.

If γ is not an even positive integer, then Corollary 3 cannot be applicable since in that case the function

$$|\xi|^{\gamma}(1+\rho(\xi))$$

can of course never satisfy the differentiability assumption of Corollary 3. The following result, however, is sometimes applicable in that case.

Corollary 5. Assume that $K_t(x) = t^{-n} k(x/t)$, $k \in L_1(R^n)$, and that $\hat{k}(\xi)$ is of the form

$$1-\hat{k}(\xi) = F(|\xi|^{\gamma})$$

in some neighborhood of the origin, where γ is a real number > 0 and F is a function of one variable of class $C^{[n/2]+2}$ such that $F(0) = 0$ and $F'(0) \neq 0$. Then the assertions A and B of Corollary 2 are valid.

Proof. It follows from the assumptions that for ξ in some neighborhood of the origin

$$1-\hat{k}(\xi) = c|\xi|^{\gamma}(1+H(|\xi|^{\gamma})) ,$$

where $c \neq 0$ and $H \in C^{[n/2]+1}(R)$, $H(0) = 0$. We claim that the function $\xi \to H(|\xi|^{\gamma})$ belongs to $\hat{M}(R^n)$ locally in some neighborhood of the origin. The assertion then follows from Corollary 2. In fact one easily shows that an arbitrary derivative of degree m of the function $\xi \to H(|\xi|^{\gamma})$ is $O(|\xi|^{\gamma-m})$ as $|\xi| \to 0$, so all we have to do is to apply Lemma 1.

4. Proof of the theorem. If E is any Banach space and E′ its dual space, then the topology $\sigma(E', E)$ on E′ (sometimes called the weak* topology) is the topology defined by all the linear forms $E' \to \overline{C}$

$$u \to u(y) ,$$

where y runs through E. Here u(y) denotes the value at y of the linear form $u \in E'$. Now it is well known that for any separable Banach space E, the closed unit ball in E′ is sequentially compact in the topology $\sigma(E', E)$. Taking $E = L_p$ we obtain $E' = L_{p'}$ if $1 \leq p < \infty$ and $p'^{-1} + p^{-1} = 1$. Choosing for E the space C_0 of continuous functions tending to zero at infinity equipped with the usual norm, we get E′ = M. Clearly, in both these cases the $\mathcal{S}'$-topology is weaker than the topology $\sigma(E', E)$. Replacing p′ by p we thus obtain the following statement.

<u>Proposition 1</u>. <u>The closed unit balls in</u> L_p, $1 < p \leq \infty$, <u>and</u> M <u>are sequentially compact in the</u> $\mathcal{S}'$-<u>topology</u>.

As is well known the unit ball in L_1 is not sequentially compact (and not compact) in the $\mathcal{S}'$-topology.

<u>Proof of A</u>. From the assumptions (3.1) and (3.2) it follows immediately that

$$\alpha(t)^{-1}(\hat{K}(t, \xi) - 1) \xrightarrow{\mathcal{S}'} q(\xi), \text{ as } t \to 0 .$$

Here $\xrightarrow{\mathcal{S}'}$ denotes convergence in the $\mathcal{S}'$-topology. Next, take the inverse Fourier transforms (in the distribution sense) of both members, and form convolutions with f. Since both these operations are continuous in the $\mathcal{S}'$-topology, we obtain at once

$$(4.1) \qquad \alpha(t)^{-1}(W(t, \cdot, f) - f(\cdot)) \xrightarrow{\mathcal{S}'} Q * f, \text{ as } t \to 0 .$$

But the assumption that the approximation is of the order $o(\alpha(t))$ as $t \to 0$ implies of course that

$$(4.2) \qquad \alpha(t)^{-1}(W(t, \cdot, f) - f(\cdot)) \xrightarrow{\mathcal{S}'} 0, \text{ as } t \to 0 ,$$

since the topology of $\mathcal{S}'$ is weaker than that of L_p. Combining (4.1) and (4.2) we obtain $Q * f = 0$.

<u>Proof of B</u>. By Proposition 1 it follows from condition (i) of the theorem that there exists a sequence $t_k \to 0$ and an element $g \in L_p$ $(1 < p \leq \infty)$, or $g \in M$ $(p = 1)$, respectively, such that

$$\alpha(t_k)^{-1}(W(t_k, \cdot, f) - f(\cdot)) \xrightarrow{\mathcal{S}'} g, \text{ as } k \to \infty .$$

Combining this with (4.1) we conclude that g must be equal to $Q * f$, and hence that $Q * f \in L_p$ if $1 < p \leq \infty$ and $Q * f \in M$ if $p = 1$.

Conversely, assume that $f \in L_p$ and $Q * f \in L_p$. Let R_t be the measure whose Fourier transform is $\rho(t, \xi)$. Then we obtain from (3.1) and (3.2)

$$\|\alpha(t)^{-1}(W(t, \cdot, f) - f(\cdot))\|_p \leq \|Q * f\|_p + \|Q * R_t * f\|_p \leq (1 + B)\| Q * f\|_p ,$$

which proves (ii) $\Longrightarrow$ (i) in the case $1 < p \leq \infty$. The proof for the case $p = 1$ is similar. This completes the proof of the theorem.

5. Examples. (1) Almost all the kernels considered in the literature, for instance by Butzer, Nessel, Görlich, and by Kozima and Sunouchi in the cited articles, satisfy the assumptions of Corollary 4. So do for instance the generalized Gauss-Weierstrass kernel

$$k(x) = ce^{-|x|^{\lambda}} , \tag{5.1}$$

where $\lambda > 0$ and c is chosen so that $\int k(x)\,dx = 1$, the mean value kernels M_1 and S_1, as well as the iterates of these kernels. The mean value kernels M_1 and S_1 are defined — up to a multiplicative constant — as the characteristic function for the unit ball in R^n, and the surface measure on the unit sphere in R^n, respectively. Hence the approximation by means of any of these kernels has the saturation properties described in Corollary 2 for $\gamma = 2$. Specializing the parameter λ in (5.1) to $\lambda = 2$ and $\lambda = 1$ we obtain the ordinary Gauss-Weierstrass kernel and the Picard kernel, respectively.

(2) A generalized Riesz kernel can be defined for $\gamma > 0$ and $\alpha > (n - 1)/2$ by

$$\hat{k}(\xi) = (\max (0, 1 - |\xi|^{\gamma}))^{\alpha}$$

It is known that $\hat{k} \in \hat{L}_1$ for the given values of the parameters α and γ (for a proof of this fact see e.g. the already cited article by Löfström [14]). Here we may apply Corollary 5. In fact it is sufficient to note that

$$1 - \hat{k}(\xi) = F(|\xi|^{\gamma}) ,$$

where

$$F(u) = 1 - (1 - u)^{\alpha}$$

for u in a neighborhood of the origin. Hence we obtain saturation of order γ in this case. The saturation class is again determined by Corollary 2.

In the case of the classical Riesz kernel, which corresponds to $\gamma = 2$, $\hat{k}(\xi)$ is infinitely differentiable in a whole neighborhood of the origin, so that in this case we can, if we wish, directly apply Corollary 3.

(3) The Cauchy-Poisson kernel is defined by

$$k(x) = c(1 + |x|^2)^{-(n+1)/2}, \ x \in R^n ,$$

where c is chosen so that $\int k(x)\,dx = 1$. Here we cannot use Corollary 4, since k(x) does not decrease rapidly enough at infinity. Instead we use the well-known fact that

$$\hat{k}(\xi) = e^{-|\xi|}, \ \xi \in R^n . \tag{5.2}$$

It follows from (5.2) and Corollary 5 that in this case the assertions A and B of Corollary 2 are valid with $\gamma = 1$.

References

1. Buchwalter, H.: Saturation et distribution, C.R. Acad. Sci. Paris 250 (1960), 3562-3564.

2. Burenkov, V.I.: Imbedding and continuation for classes of differentiable functions of several variables defined in the whole space, Progress in Mathematics (ed. R.V. Gamkrelidze) vol. 2 (1968), 73-161.

3. Butzer, P.L. and Nessel, R.J.: Contributions to the theory of saturation for singular integrals in several variables. I, Indag. Math. 28 (1966), 515-531.

4. Nessel, R.J.: Contributions to the theory of saturation ... II, Indag. Math. 29 (1967), 52-64.

5. Nessel, R.J., Contributions to the theory of saturation ... III, Indag. Math. 29 (1967), 65-73.

6. Berens, H. and Nessel, R.J.: Contributions to the theory of saturation ... IV, Indag. Math. 30 (1968), 325-335.

7. Berens, H. and Nessel, R.J.: Contributions to the theory of saturation ... V, Indag. Math. 31 (1969), 71-88.

8. Görlich, E.: Distributionentheoretische Methoden in der Saturationstheorie. Dissertation, Aachen 1967.

9. Görlich, E.: Distributional methods in saturation theory, J. Approx. Theory 1 (1968), 111-136.

10. Görlich, E.: Saturation theorems and distributional methods. Abstract spaces and Approximation, Proc.Conf. Oberwolfach, Birkhäuser Verlag 1969, 218-232.

11. Hörmander, L.: Estimates for translation invariant operators in L_p-spaces, Acta Math. 104 (1960), 93-140.

12. Kozima, M. and Sunouchi, G.: On the approximation and saturation by general singular integrals, Tôhoku Math. J. 20 (1968), 146-169.

13. de Leuuw, K. and Mirkil, H.: A priori estimates for differential operators in L_∞-norm, Illinois J. Math. 8 (1964), 112-124.

14. Löfström, J.: Some theorems on interpolation spaces with applications to approximation in L_p, Math. Ann. 172 (1967), 176-196.

15. Schwartz, L.: Théorie des distributions, I, II, Hermann, Paris 1950-51.

16. Sunouchi, G.: On the class of saturation in the theory of approximation I, Tôhoku Math. J. 12 (1960), 339-344.

APPENDIX II.

The Kolmogorov Superposition Theorem

by Torbjörn Hedberg

1. Introduction. The purpose of this note is to give a simple proof of a theorem, due essentially to Kolmogorov, concerning the representation of continuous functions of any number of variables as "superpositions" (in the sense of functional composition) of functions of only one and two variables. The theorem in question has its origin in the famous collection of problems which Hilbert submitted to the International Congress of Mathematicians in Paris, 1900 (Hilbert [5]). In enunciating the thirteenth of these problems, Hilbert conjectured that a certain quite specific continuous (in fact, analytic) function of three variables, arising in the solution of the general equation of degree seven, is not representable by means of composition of functions of one and two variables.

This matter remained unsettled until the late 'fifties. In 1956, Kolmogorov [7] proved the surprising result that every continuous function, of an arbitrary number of variables, is representable as a composition of functions of three (or fewer) variables. In 1957 his student Arnold showed that "three" can be replaced by "two" (Arnold [1]). Soon after, Kolmogorov [8] showed that every continuous function f on the unit n-cube admits a representation

$$(*) \qquad f(x_1, \ldots, x_n) = \sum_{i=1}^{2n+1} g_i\Big(\sum_{j=1}^{n} \varphi_{ij}(x_j)\Big)$$

where the g_i and φ_{ij} are continuous, and moreover the φ_{ij} are determined in advance, independently of the choice of f. This, of course, disproves Hilbert's conjecture in a very strong form, representing f as a superposition of continuous functions of <u>one</u> variable, and of the single function $(x, y) \to x + y$ of two variables. Later, various mathematicians further refined this result, to give it the form in which we shall enunciate it below in Theorems 1 and 2 (for references to these works, as well as an authoritative account of the developments initiated by Hilbert's thirteenth problem, see Vitushkin [10]).

The proof we shall give of the superposition theorem is based on that in Lorentz ([9], Ch. 11), and on work of Kahane dealing with a closely related problem (see Kahane [6], pp. 98 - 101). The idea of using a Baire category argument is due to Kahane. This idea simplifies the proof, and is especially useful for obtaining various refinements of the result.

2. The superposition theorem. We denote by I the closed interval $[0, 1]$, and by $C(I^n)$, $n \geq 1$, the real-valued continuous functions on I^n, made into a Banach space in the usual way. By $\|f\|$ we always denote the supremum of $|f|$.

Theorem 1. *If $n \geq 2$, there exist real numbers $\lambda_1, \dots, \lambda_n$ and elements $\varphi_1, \dots, \varphi_{2n+1} \in C(I)$ with the following property: for each $f \in C(I^n)$ there exists $g \in C(R)$ such that*

$$(1) \qquad f(x_1, \dots, x_n) = \sum_{k=1}^{2n+1} g(\lambda_1 \varphi_k(x_1) + \dots + \lambda_n \varphi_k(x_n)) .$$

Remarks. The proof will actually show more, namely $\lambda_1, \dots, \lambda_n$ can be so chosen that all $(2n+1)$-tuples $(\varphi_1, \dots, \varphi_{2n+1})$, except for a set of first category (relative to the metric space $C(I)^{2n+1}$) have the stated property. Afterwards we shall discuss variants in which the φ_i are required to be monotone, Hölder continuous, etc.

The proof of Theorem 1 will be carried out in the case $n = 2$, for notational simplicity. The general case is done exactly the same way, except that further discussion will be needed regarding Lemma 1. Q shall denote the set of rational numbers.

Lemma 1. *There exists a real number λ such that $x, y, x', y' \in Q$, $x + \lambda y = x' + \lambda y'$ imply $x = x'$, $y = y'$.*

Proof. Choose for λ any irrational number.

The heart of the theorem is contained in the next lemma.

Lemma 2. *Let λ be a (throughout, fixed!) number satisfying the hypothesis of Lemma 1*

Let $f \in C(I^2)$, $\|f\| = 1$, and let U_f denote that subset of $C(I)^5$ described as follows. The 5-tuple $(\varphi_1, \ldots, \varphi_5)$ belongs to U_f if and only if there exists $g \in C(R)$ such that

(i) $$|g(t)| \leqslant 1/7 , \quad t \in R$$

(ii) $$|f(x, y) - \sum_{i=1}^{5} g(\varphi_i(x) + \lambda\, \varphi_i(y))| < 7/8 , \quad (x, y) \in I^2 .$$

Then, U_f is an open dense subset of $C(I)^5$.

Proof. The set U_f is open, since if g "works" for a particular 5-tuple, it does so for all sufficiently neighboring 5-tuples. We'll now show that U_f is dense in $C(I)^5$ (the latter being considered as a metric space in the natural way, i.e. the distance between $(\varphi_1, \ldots, \varphi_5)$ and $(\varphi_1', \ldots, \varphi_5')$ is $(\sum_{i=1}^{5} \|\varphi_i - \varphi_i'\|^2)^{1/2}$).

Let $\varphi_1^o, \ldots, \varphi_5^o$ be any given functions in $C(I)$, and $f \in C(I^2)$, $\|f\| = 1$. We must show, given $\varepsilon > 0$, we can find $\varphi_i \in C(I)$ and $g \in C(R)$ such that $\|\varphi_i - \varphi_i^o\| < \varepsilon$ and (i), (ii) hold. Let N denote a positive integer (to be specified later). If i is one of the integers 1, 2, 3, 4, 5 consider the set of sub-intervals of I which remain when all of the intervals $[s/N, (s + 1)/N]$ with $0 \leqslant s < N$, $s \equiv i - 1 \pmod 5$ are deleted. These remaining intervals, with end-points adjoined so that they are closed, shall be designated the red intervals of rank i. Each red interval of rank i has length $4/N$, with two possible exceptions.

Now, if N is large enough, it is easy to verify that functions $\varphi_i \in C(I)$ exist with the following properties:

a) φ_i is constant, equal to a rational number, on each red interval of rank i.

b) φ_i takes distinct values on distinct red intervals of rank i; moreover, if $i \neq j$, the value φ_i takes on any red interval of rank i differs from the value which φ_j takes on any red interval of rank j.

c) $\|\varphi_i - \varphi_i^o\| < \varepsilon , \quad i = 1, \ldots, 5 .$

A rectangle lying in I^2 which is the Cartesian product of two red intervals of rank i (one lying in $\{0 \leqslant x \leqslant 1\}$, one lying in $\{0 \leqslant y \leqslant 1\}$) will be called a red

<u>rectangle</u> of rank i (nearly all of these will of course be squares, of side $4/N$). The red rectangles of rank i shall be designated $R_{i,1}$, $R_{i,2}$, ... Let us define

$$\Phi_i(x, y) = \varphi_i(x) + \lambda\, \varphi_i(y) \ , \quad i = 1, \ldots, 5 \ . \tag{2}$$

Observe that Φ_i is constant on each red rectangle of rank i, and by virtue of Lemma 1, <u>the (constant) value, denoted by $\Phi_i(R_{i,r})$, which Φ_i takes on $R_{i,r}$, cannot equal the (constant) value which Φ_j takes on $R_{j,s}$ except trivially, that is if</u> $i = j$, $r = s$. Finally, let us suppose N chosen so large that

$$|f(x, y) - f(x', y')| \leq 1/7 \quad \text{whenever} \quad (x - x')^2 + (y - y')^2 \leq 32/N^2 \tag{3}$$

which is evidently possible by uniform continuity.

We shall now construct the function g. Let us examine in turn each of the red rectangles (of all ranks); consider some particular one, $R_{i,r}$. Then,

<u>if</u> $f(x, y) > 0$ <u>throughout</u> $R_{i,r}$, <u>we</u>
<u>define</u> $g(\Phi_i(R_{i,r})) = 1/7$;

<u>if</u> $f(x, y) < 0$ <u>throughout</u> $R_{i,r}$, <u>we</u>
<u>define</u> $g(\Phi_i(R_{i,r})) = -1/7$.

Because the numbers $\Phi_i(R_{i,r})$ corresponding to distinct pairs (i, r) are necessarily distinct, this prescription cannot lead to any inconsistencies. Let now g be extended to the whole line (e.g. as a piecewise linear function) so that (i) holds. We shall now verify that (ii) is satisfied. Choose any point $(x, y) \in I^2$. Since x lies in red intervals of rank i, except perhaps for one value of i, and the same is true of y, (x, y) is contained in red rectangles of at least three different ranks. We consider now three cases:

If $f(x, y) > 1/7$, then because of (3), $f(x, y)$ is positive throughout each red rectangle containing (x, y), hence

$$f(x, y) - \sum_{i=1}^{5} g(\Phi_i(x, y)) \leq 1 - (3/7) + (2/7) < 7/8 \ , \tag{4}$$

since at least three of the numbers $g(\Phi_i(x, y))$ are equal to 1/7. Also, the left side of (4) is larger than $(1/7) - (5/7) = - 4/7$, so that (ii) is verified in this case. The case $f(x, y) < - 1/7$ is treated similarly.

Finally, if $|f(x, y)| \leq 1/7$, the expression on the left of (4) has absolute value not exceeding $6/7$, so (ii) holds also in this case, and the lemma is proved.

Lemma 3. With λ as in Lemma 2, there exist $\varphi_1, \ldots, \varphi_5 \in C(I)$ such that, given $f \in C(I^2)$, there exists $g \in C(R)$ satisfying

(i) $$|g(t)| \leq (1/7) \|f\| , \quad t \in R$$

(ii) $$\|f - \sum_{i=1}^{5} g \circ \Phi_i\| \leq (8/9) \|f\| .$$

Here Φ_i are the functions defined by (2).

Proof. Clearly it is no loss of generality to assume $\|f\| = 1$. Let $\{h_j\}_{j=1}^{\infty}$ be a sequence of functions in $C(I^2)$, each having norm 1, which is dense in the unit sphere of $C(I^2)$. Let $U_j = U_{h_j}$ denote the subset of $C(I)^5$ determined by h_j, as in the preceding lemma. Since the U_j are dense open subsets of the complete metric space $C(I)^5$, their intersection V is non-empty (indeed, a dense set of second category) by Baire's theorem. Choose $(\varphi_1, \ldots, \varphi_5) \in V$; we shall show that these φ_i satisfy the requirements.

Let f be any function in $C(I^2)$. We can then find m such that $\|f - h_m\| \leq 1/72$. By Lemma 2, since $(\varphi_1, \ldots, \varphi_5) \in V \subset U_m$, there exists $g \in C(R)$, $\|g\| \leq 1/7$, such that $\|h_m - \sum_{i=1}^{5} g \circ \Phi_i\| < 7/8$. Hence

$$\|f - \sum_{i=1}^{5} g \circ \Phi_i\| \leq (7/8) + (1/72) = 8/9 ,$$

proving the lemma.

Proof of Theorem 1 (case n = 2). Choose $\lambda_1 = 1$, $\lambda_2 = \lambda$ and let $(\varphi_1, \ldots, \varphi_5)$ be as in Lemma 3. We shall show that the requirements of Theorem 1 are satisfied. Given $f \in C(I^2)$, we define by recursion a sequence $\{f_j\}_{j=0}^{\infty}$ in $C(I^2)$, and a sequence $\{g_j\}_{j=0}^{\infty}$ in $C(R)$, as follows:

a) $$f_o = f$$

b) $$f_{j+1} = f_j - \sum_{i=1}^{5} g_j \circ \Phi_i$$

where $g_j \in C(R)$ is chosen so that

$$\|g_j\| \leq (1/7)\ \|f_j\|\ ,\quad \|f_j - \sum_{i=1}^{5} g_j \circ \Phi_i\| \leq (8/9)\ \|f_j\|$$

(such g_j exist, by Lemma 3).

Then $\|f_{j+1}\| \leq (8/9)\ \|f_j\|$, hence $\|f_j\| \leq (8/9)^j\ \|f\|$ and $\|g_j\| \leq (1/7)(8/9)^j\ \|f\|$. Hence the series $\sum_{j=o}^{\infty} g_j$ converges in norm to an element $g \in C(R)$, and we have

$$f = \sum_{j=o}^{\infty} (f_j - f_{j+1}) = \sum_{j=o}^{\infty} \sum_{i=1}^{5} g_j \circ \Phi_i = \sum_{i=1}^{5} g \circ \Phi_i\ ,$$

completing the proof.

Remark 1. To do the general case, Lemma 1 must be extended as follows:

Lemma 1'. There exist real numbers $\lambda_1, \ldots, \lambda_n$ such that $x_1, x_1', \ldots, x_n, x_n' \in Q$ and

(5) $$\lambda_1 x_1 + \ldots + \lambda_n x_n = \lambda_1 x_1' + \ldots + \lambda_n x_n'$$

imply

(6) $$x_i = x_i'\ ,\quad i = 1, 2, \ldots, n\ .$$

Proof. We have only to choose the λ_i so that they are independent over the rational field, e.g. $\lambda_i = \pi^{i-1}$, $i = 1, 2, \ldots, n$.

An alternative proof is to observe that the set of $(\lambda_1, \ldots, \lambda_n) \in R^n$ for which (5) holds for given x_i, x_i' while (6) does not, is a hyperplane. Hence all n-tuples, apart from the points of a certain countable union of hyperplanes (which is a set of measure zero and first category in R^n) satisfy the requirements.

Remark 2. We can modify the proof of Theorem 1 so as to obtain φ_i which are non-

<u>decreasing</u>. Simply let H denote the closed subset of C(I) consisting of non-decreasing functions, and use the Baire category argument in H^5 rather than $C(I)^5$. We can also, although this is less obvious, arrange that the φ_i shall be of class Lip α, for any $\alpha < 1$, and even somewhat more. This is discussed in the next section.

<u>3. A stronger form of the superposition theorem.</u> It seems likely from work of B.L. Fridman [3] that the φ_i in Theorem 1 could be chosen to be of class Lip 1 on [0, 1]. (Specifically, Fridman works with the Kolmogorov representation (*) rather than (1), and proves that the φ_{ij} may be chosen in Lip 1.) This would be essentially unimprovable, since general results of Vitushkin and Henkin (see [10], p. 168, for references) imply that continuously differentiable φ_i satisfying the requirements of the theorem cannot exist.

The "category" argument employed in the preceding section can be modified to yield a refinement of Theorem 1, whereby the φ_i are "nearly" of class Lip 1. In order to formulate this result, let h denote a positive non-decreasing subadditive function on $[0, \infty)$ which satisfies

$$\lim_{t \to 0^+} h(t)/t = \infty$$

Let H_h denote the set of all real-valued $\varphi \in C(I)$ satisfying

$$\omega(\varphi, t) = o(h(t)), \quad t \to 0 \tag{1}$$

for which

$$\|\varphi\|_h \underset{\text{def.}}{=} \|\varphi\|_\infty + \sup_{t > 0} [\omega(\varphi, t)/h(t)] \tag{2}$$

is finite, whereby $\omega(\varphi, t)$ denotes the modulus of continuity of φ. It is easy to verify that H_h is a Banach space with respect to the norm (2). With these notations we have

<u>Theorem 2.</u> <u>The</u> φ_i <u>in Theorem</u> 1 <u>can be chosen from</u> H_h. <u>In fact,</u> $\lambda_1, \ldots, \lambda_n$ <u>can be so chosen that all</u> $(2n + 1)$-<u>tuples</u> $(\varphi_1, \ldots, \varphi_{2n+1})$, <u>except for a set of first category (relative to the metric space</u> H_h^{2n+1}) <u>have the stated property</u>.

The proof we have given for Theorem 1 goes through in all details also in this case. The only point which requires further discussion is that which underlies property c) of the φ_i in the proof of Lemma 2; that is, one must establish

Lemma 4. Let $\varphi \in H_h$ and $\epsilon > 0$. Then, there exists N_0 such that, for each subdivision of I into intervals

$$r_{N,k} = [k/N, (k+1)/N], \quad k = 0, 1, \ldots, N-1 ,$$

with $N \geq N_0$, there exists a function ψ in H_h, linear on $r_{N,k}$ for $k \equiv 1 \pmod 5$ and constant elsewhere, such that $\|\psi - \varphi\|_h \leq \epsilon$. (The corresponding assertions for the remaining residue classes mod 5 are also valid).

We omit the simple proof, which is similar to that of Lemme 1 on p. 99 of [6].

Just as with Theorem 1, we can also arrange that the φ_i in Theorem 2 be non-decreasing, by working from the outset with the metric space of non-decreasing elements of H_h rather than H_h itself.

4. Concluding remarks.

a) Theorems 1 and 2 remain valid even if g is restricted to belong to certain subclasses of $C(R)$. Suppose, to fix ideas, that the functions Φ_i have their ranges in $[0, c]$ where $c < 2\pi$. Then (cf. [4]) one can choose $g(t) = f(e^{it})$, where f is an element of the "disc algebra", i.e. f is continuous on $\{|z| \leq 1\}$ and analytic on $\{|z| < 1\}$.

b) Doss [2] has proved that, in the case $n = 2$, the number of summands (i.e. five) in (1) cannot be reduced. Whether the number $2n + 1$ is best possible also when $n > 2$ is not known.

References

1. V.I. Arnold, On functions of three variables, Dokl. Akad. Nauk SSSR 114 No. 4 (1957) 679-681 (Russian). (Amer. Math. Soc. Transl. (2) 28 (1963) 51-54).
2. R. Doss, On the representation of continuous functions of two variables by means of addition and continuous functions of one variable, Colloquium Math. 10 (1963) 249-259.
3. B.L. Fridman, An improvement in the smoothness of the functions in A.N. Kolmogorov's theorem on superpositions, Dokl. Akad. Nauk SSSR 177 (1967) 1019-1022 (Russian).
4. T. Hedberg, Sur les réarrangements de fonctions de la classe A et les ensembles d'interpolation pour $A(D^2)$, Comptes Rendus, 270, série A (1970) 1491-1494.
5. D. Hilbert, Mathematische Probleme, Gesamm. Abh. III, 290-329.
6. J.-P. Kahane, Séries de Fourier absolument convergentes, Erg. der Math. Band 50, Springer-Verlag, 1970.
7. A.N. Kolmogorov, On the representation of continuous functions of several variables by means of superpositions of continuous functions of a smaller number of variables, Dokl. Adad. Nauk SSSR 108, No. 2 (1956) 179-182 (Russian).
8. ———, On the representation of continuous functions of several variables in the form of a superposition of continuous functions of one variable and addition, Dokl. Akad. Nauk SSSR 114 No. 5 (1957) 953-956 (Russian). (Amer. Math. Soc. Transl. (2) 28 (1963) 55-59).
9. G.G. Lorentz, Approximation of Functions, Holt, Rinehart and Winston 1966.
10. A.G. Vitushkin, On Hilbert's thirteenth problem, in "Hilbert's Problems", S.S. Demidov ed., Nauka, Moscow 1969, pp. 163-170 (Russian).

